Farm Animals for Gardens and Backyards

Ann Cliff

Illustrations by Neville Cliff

Aird Books
MELBOURNE

Aird Books Pty Ltd
PO Box 122
Flemington, Vic.3031
Phone (03) 376 4461

First published by Aird Books in 1993

Copyright © Ann Cliff 1993

National Library of Australia
Cataloguing-in-publication data

Cliff, Ann.
  Farm animals for gardens and backyards.

  Includes index.
  ISBN 0 947214 45 3.

  1. Self-sufficiency – Australia. 2. Domestic animals –
Australia. 3.
  Livestock – Australia. I. Title.

636.00994

All rights reserved. No part of this publication may be
reproduced, stored in a retrieval system, or transmitted in
any form or by any means, electronic, mechanical,
photocopying, recording, or otherwise, without the written
permission of the publishers.

Cover design and illustration by Lin Tobias
Text illustrations by Neville Cliff
Typeset in-house by Aird Books
Printed by Australian Print Group, Maryborough, Victoria

Contents

1 People and Animals   1
Success with animals   3
The environment   4

2 Animal Welfare   6
The RSPCA   8
The Australian Federation for the Welfare of Animals   9
Keeping animals happy   10
Alternative medicine for animals   13
    Nursing – Herbal medicine –
    Homoeopathic medicine – Probiotics
Disposal of farm animals   15

3 The Backyard Farmyard   17
This business of self-sufficiency   17
Help, advice and training   17
Equipment   20
Managing the block   21
Milk handling and milk products   27
Fibre from goats, sheep and rabbits   34
Meat production   37

4 The Animals

Bees	40		Hens	92
Cows	51		Horses	104
Dogs	61		Pigs	110
Ducks	68		Rabbits	121
Fish	74		Sheep	126
Geese	79		Worms	134
Goats	82			

Bibliography 138

Index 139

1

People and Animals

The company of animals is good for us. There are many stories about the therapeutic effects of keeping pets. Merely stroking a dog or a cat can relieve tension in a person. Distressed and lonely people become healthier and happier with an animal to care for. Children also gain a lot from keeping animals: a pet gives them companionship and affection. They learn biology and responsibility, as well as having fun.

All we need for rapport with animals is to like them. Unless we do, we shall not enjoy keeping them. People who are successful with livestock are those who take a great interest in their animals. They have the ability to empathise with them, and the animals learn to trust the people who have their welfare at heart.

If you are thinking of keeping an unusual animal, find out as much as you can about them before making up your mind. Above all, meet and mingle with them so that you interact in a practical way. Most people can get on with some kinds of animal, but few are good with all animals.

Smell can put you off a particular kind of animal. Pigs, for example, have a very persistent smell, which is not really unpleasant outdoors, but could be a drawback if you carry the smell indoors. The smell can be hard to get rid of if you have to feed and clean out pigsties before you go to work.

Fortunately for us, animals tolerate us very well, in their different ways. 'Dogs look up to you' goes the old saying, 'cats look down on you, but pigs is equal.' Goats will adopt us as herd members, and cows will give us a place in their hierarchy – but not always at the top, in my experience!

About forty species of animals are domesticated, and living in symbiosis with people. Some animals (cats and llamas, for example) are rather aloof, and give the impression of staying with us only because it suits them.

Animals most 'developed' by farmers differ the most from their wild ancestors. The small and hardy primitive breeds of sheep can forage for themselves and lead almost self-sufficient lives, but the modern European breeds (such as the Texel) will probably need housing in winter and assistance at lambing. Shearing will be essential to rid them of their heavy wool in early summer, whereas a wild sheep sheds its wool naturally in spring.

In the case of horses and cows, the wild ancestor is now extinct, having disappeared in the 6 000 or so years since the process of domestication began. There are four kinds of wild sheep left, but the Bighorn of North America is probably the only one that has made no genetic contribution to the domestic breeds of sheep.

Whether we are sitting by the fire with a cat or riding through the bush on a horse, it is interesting to know the origins of our companion, how it lived in the wild and how evolution fitted it for a particular niche in a certain ecosystem. The more we learn, the more we can understand the behaviour of animals and predict their reactions. Our understanding naturally leads to a better life for the animals – and for us as well, since they will respond to our thoughtful treatment of them.

Much of the bad behaviour in animals is the result of poor handling. For example, when a horse associates transport with an unpleasant experience, and it has come to dislike a float, the horse may refuse to get into it, and a horse is too big to move against its will. So the experiences we give our animals are crucial to their view of the world and thus to their co-operation with us.

Most of the animals we live with are gregarious in the wild, living in social groups with a definite hierarchy. Domestication has sometimes disrupted this order, especially if the animals are kepts indoors. But instincts survive, and if we give a group of caged hens their freedom, they will form themselves into a flock and behave in a manner similar to that of their wild ancestors.

It is always best to keep animals with others of their own kind, because humans are no substitute for another ferret or sheep. In fact, sheep need several others around before they feel really comfortable.

The desire to be with animals runs very deep with some humans. Although the aims of farmers appear to be purely commercial, they take great pride in their healthy stock, know them as individuals, and are concerned when they are sick.

Dairy cows that come into the shed twice a day for milking are rather like colleagues: we like some better than others, but we have some sort of relationship with them all. They know us and they greet us after we've had a weekend off. Even though cows don't jump about like dogs, they will give us a long look from their large placid eyes. When strangers turn up at milking time, cows are sometimes so wary that they won't let down their milk – a point to remember when milking a goat or a cow.

The dog was probably the first creature to come to live with humankind, and it is thought that the dingo was brought to Australia by prehistoric people. Australia's contribution to the list of domesticated species is the budgerigar, but that was comparatively recently.

We owe a lot of human progress to the animals. Dogs have made it possible to do many things more efficiently: to hunt, to herd cattle and sheep, and to protect us from our enemies are all extensions of a dog's natural instincts. Draught animals (such as oxen, horses and donkeys) made it possible in the past for large numbers of people to migrate long distances and to develop agriculture wherever they settled.

Some people like unusual pets: I know of people who walk ferrets on a lead in the park. But beware of trying to make house pets of unusual species! Animals that are not born in a nest are not amenable to house-training. Grazing animals, such as goats and sheep, scatter their pellets everywhere and have no instinct to keep a bed clean. Of the ruminants, only llamas and alpacas seem to have tidy habits, always returning to deposit their dung in the same place.

Symbiosis is the biologist's word for the living together of organisms, and it implies benefits to both sides. We get enormous benefits from our association with animals: food, transport, clothing, recreation, security, pest control and companionship. What do we provide in return? We have obligations to the animals we keep. And the more control we have over their environment, the greater our responsibility.

In the past, we struck a sort of balance between our needs and those of the animals. But things have changed in the last fifty years, with the development of large livestock units, commonly called 'factory farms'. When thousands of hens or hundreds of pigs are kept in a group, it is not possible to know the stock. In such circumstances, it usually is not possible to allow the animals much opportunity to behave naturally.

There has been much criticism of the treatment of animals in larger livestock units. Codes of practice have been drawn up in most countries, to remind people of their obligations. The codes set out the minimum standards thought to be acceptable for livestock comfort. Everyone should be familiar with the welfare codes. They may have been framed with intensive units in mind, but they apply to backyard livestock too.

Success with animals

To keep animals successfully, the motto is: *be prepared*. There are many ways in which we can learn about the species we have chosen. For example, we can:
- Read about them.
- Join a club. Look out for goat societies and smallholders' clubs, for example. (Local libraries will have their addresses.)
- Ask the vet. Get to know a good vet before you need one.
- Talk to other owners. Most will be happy to share their knowledge.
- Try to get some experience in handling the animals, like milking a goat, mucking out stables, feeding or cleaning an animal, or helping condition-scoring sheep, and so on.
- Join WWOOF (Willing Workers on Organic Farms, see Chapter 3).

Having done this, it will be easier to prepare living quarters, work out where to get food, find a male of the species for breeding, fulfil their welfare needs or take any other practical steps.

Once the animals have arrived safely, the recipe for success lies mainly in the time we spend with the animals, our powers of observation, and the way we behave towards them. The basic rules are the same for all species of domestic livestock.

Some simple rules

Give the animals peace and quiet to settle down when they arrive. Make sure there is plenty of food and water available and that there aren't too many strangers peering at them or loud noises near them. Keep the children and their friends away for a while.
- Make changes gradually. Moving from one holding to another will have been a big change for the animals. To minimise the shock, provide the same food their previous owners gave them, at least for a while.
- Handle the animals regularly so they approach you without fear. Treat them gently, but firmly, and do not let them behave badly.
- Watch them carefully, and note their normal behaviour. Any abnormal behaviour is usually an indication of some kind of trouble.
- Be predictable, calm and friendly. Avoid sudden movements.
- Talk to the animals. Let them see you before you touch them. Of course, tone of voice means a lot to an animal, where actual words may not.
- Do not let small children handle animals. It is too much to expect of the stock, and it can be dangerous for the children.
- Do not handle animals when they are eating, mating or giving birth, unless you have to.
- Watch out for stress. For example, domestic animals are mainly from temperate climates and may be severely affected by heat. A cool water spray is the best emergency treatment, but try to arrange their living quarters so that there is plenty of shade and coolness.
- Routine is important, especially with respect to feeding and milking times.

The environment

Animals have been kept by peasants all over the world for thousands of years. Peasant farming was sustainable, because waste material was returned to the soil. This maintained the soil's fertility so that crops could continue to be grown. For example, the soil of China has supported a very high population for centuries, because families recycled every scrap of organic material.

In so-called 'developed' countries, most people have lost the close contact with animals and with the soil. Farming is the occupation of only a small minority of people in industrialised countries.

Most people do not wish to return to the treadmill of total self-sufficiency. It is too hard and too limiting, and most of us are thankful that we have leisure time to pursue our interests instead of having to grow all our own food and clothing. But many of us feel instinctively that something has been lost, so we keep pets to maintain the link with animals, or cultivate a garden to keep us in touch with the earth.

In times of stress, such as wars or economic depressions, a vegetable garden is attractive and useful, and a few hens at the bottom of the garden or a hutch of rabbits have seen families through the hard times before. But an interest in livestock is not limited to subsistence farming. It's more than a means of survival.

People of modest affluence may run a hobby farm or a large garden as a family enterprise in which all members learn together. On a small area of land it is possible to keep a few animals for pleasure, for production, or for showing.

It is sometimes hard to get a balanced view of keeping livestock because animal enthusiasts are so very enthusiastic. All objections are swept away in a tide of exuberance as they explain in detail how impossible it is to live without a ferret. My aim is to try to balance the picture, to look at both the positive and negative sides.

Regulations about animal keeping protect us from our neighbours, our animals from people, the population from disease and, in some cases, they guard against over-production of food. Regulations vary between local authorities and between states. For this reason it is impossible to give a comprehensive list of rules about keeping livestock. Instead, I shall use one country town as an example. With a population of about 19 000 people, its regulations reflect a need to balance residential, industry and farming interests. There are three zones in this town: the business area, the residential area, and the semi-rural area on the outskirts of the city. Naturally, the rules about keeping animals are tighter in the middle of town. The relevant law states:

Keeping Animals
No owner or occupier of property may without a permit keep or allow to be kept more than four different types of animals on any one property at any time and must not keep, without a permit or allow to be kept any more in number for each type of animal than as set out in the following table.

Type of Animal	Maximum Allowed in Area 1	Maximum Allowed in Area 2	Maximum Allowed in Area 3
Bees (hives)	0	0	3
Cattle	0	0	no maximum
Dogs	2	2	6
Fish (domestic)	no maximum	no maximum	no maximum
Goats	0	2	no maximum
Guinea pigs	2	2	10
Horses, donkeys, etc	0	2	no maximum
Poultry	10	10	20
Pigeons	0	0	20
Pigs	0	2	no maximum
Rabbits (domestic)	2	2	2
Sheep	0	2	no maximum
Any other agricultural animal	0	0	no maximum

The law says that animals must be provided with shelter and that the area where they're kept must be maintained in a clean condition. People who want to exceed the permissible numbers must apply for a permit. Factors that will be taken into account include the likely effect on and proximity to the neighbours, the amenity of the area, the zoning of the land, and the adequacy of shelters.

Where a permit is needed for bees, the rules are quite strict, requiring ten conditions to be met before a permit is granted.

2

Animal Welfare

Over the last 6 000 years, we humans have had much practice in dealing with animals and looking after their needs. Our domesticated species have evolved from wild ancestors to be efficient producers of meat, milk, fur and wool, or to be workers and companions for human kind. And people have been observing animals, living close to them and supplying them with food and shelter for many generations.

I believe two things make people good with livestock: temperament, which can be modified, and training.

Fairly quiet people often have the right temperament. They are keen observers. They watch the stock all the time out of interest. They remember what they see. In this way they build up a picture of how their particular class of stock behaves when going about its normal business.

If we know the normal patterns of animal behaviour, problems are much easier to spot. Recognising the first sign of ill health or of an approaching birth will give us time to prepare ourselves and handle the situation properly. Patience is another character trait the animal handler needs. Patience is necessary with animals because they can obstruct us when we are trying to help them, or when we have an urgent appointment elsewhere.

Deer come to mind as creatures needing enormous patience. Deer are relatively undomesticated, and their natural fear of people makes them hard to handle. Deer-farmers are quiet and persistent people, prepared to wait (sometimes for days) to complete a manoeuvre, such as moving the stock from one paddock to another.

Training is essential. Through ignorance, you can mistreat animals. Any time spent with stock is a learning experience. When you meet other people who keep the same stock, use this as an opportunity to exchange information. Experienced people will ask questions, not because they are trying to catch you out, but because they're still keen to learn more from others' experience.

Animal rights organisations are now very vocal. Do they have a point? Of course they do. There is much cruelty to animals in the world, and anyone who doubts this should visit an RSPCA animal shelter. Because animals can't speak for themselves, their interests must be safeguarded, not just by pressure groups, but by all of us.

Laws to prohibit cruelty to animals were made in the nineteenth century. Since then, things have changed. It is not only individuals, but systems of livestock-keeping that nowadays deny animals the right to behave naturally. Legislation is now being enacted in connection with intensive farming. In some countries, battery-hen cages, for example, are now illegal. The European Community has banned the installation of any new sow-stall systems (in which sows are confined without being able to turn round), and the existing ones will soon be outlawed.

Animal welfare needs more than legislation. An animal keeper must be constantly alert to conditions like extreme weather conditions, for example, that affect stock.

When we keep animals, we are responsible for the environment they live in. The more intensive the methods, the more control we have, and the more responsibility we bear.

However, where farming is extensive (the opposite of intensive 'agribusiness'), animal suffering on a large scale may be caused, particularly in extreme climates – for example, when drought hits a large sheep station.

There are several ways in which domesticated animal species can suffer: abuse, neglect and deprivation are three forms of ill-treatment. Where animals are abused or neglected, production is lowered. When they are deprived, this does not always happen.

A common defence of intensive farming practices is that when production is high, welfare must be good, so farmers will see to animal welfare in their own interests. Unfortunately for them, hens lay very well in battery cages. The profits in battery-hen systems are not large, because of the high energy inputs. So the number of birds in a cage was increased to five, even seven, in appalling, cramped conditions, where the birds can never stretch or flap their wings. But the hens keep on laying, so the system continues whilst battery eggs have a market. If enough people ask for free-range chickens and eggs, the battery cages will go.

An animal can be said to be deprived when it is prevented from behaving naturally, and intensive systems do prevent natural behaviour. Another example is the modern piggery, where there is no straw or bedding. Pigs love to arrange a bed, and when they are about to give birth, they have an urgent need to select material to make a nest. Without bedding material, this instinct is frustrated.

Bedding helps to alleviate boredom in pigs. They should also have access to the adventure of being outdoors for at least part of their lives. The relentless pacing of lions in captivity is similar to the compulsive behaviour of farm stock in modern, intensive systems.

Animal welfare recommendations provide animals with the 'five freedoms'. They are:
- Freedom from thirst, hunger and malnutrition.
- Appropriate comfort and shelter.
- Prevention or rapid diagnosis and treatment of injury and disease.

- Freedom to display most normal behaviour patterns.
- Freedom from fear.

The subject of animal rights is a matter of ethics and is hotly debated. We have moved a long way from the simple view that animals are provided for our convenience and that we have dominion over them. The position held by most people who keep animals these days is that we have a duty to use animals well and to give them a good life and a humane death.

Animals are not like humans. They do not anticipate death. However, when caught and directed against their instincts, they fear death. The whole basis of good handling is not to terrify the animal. Ruth Harrison's *Animal Machines* (Stuart, 1964) was probably the first influential book about animal welfare. Books like Peter Singer's *Animal Liberation* (Cape, 1976) shocked many people, but some of the practices described in the book are now illegal in most countries. The change in public opinion as a result of animal liberation campaigns has been considerable.

Singer is a philosopher, and his argument is that what animals need from us is not a sentimental approach, but appropriate consideration. He says that equality of species is a moral ideal, and that just as people of different gender or different race should have equal consideration, so animals as sentient beings should have equal consideration with us.

Animal welfare has been the subject of scientific research only recently. For example, a group at Nottingham University in England has studied the opinions of a herd of pigs about their handlers. The study was done on a large pig farm, where there were five identical units of 1 000 sows each. Everything was standardised, except the attendants. The researchers asked: which attendants had the happiest pigs, measured by production? One of the traits rated was the 'implied non-aggression level' of the stockperson.

Another was the way in which the person handled the pigs. Were the piglets thrown back into the pen after handling, or were they gently lowered to the ground? The researchers found that the best stockhandlers were confident introverts with a low aggression level. The report also suggested that innate ability of the stockperson mattered less than motivation. The time taken by the pigs to approach the handler depended entirely upon the person. If an animal moves confidently towards the handler, then that is most probably a good stockperson.

The RSPCA

The Royal Society for the Prevention of Cruelty to Animals is independent and community-based. In Victoria, its small government grant is actually less than its payroll tax. All its services are dependent on fundraising. It is a very popular charity, and its success in maintaining the services we take for granted is amazing. One of the main objectives of the RSPCA is education of the community with regard to the humane treatment and management of animals, through literature, lectures, and so on.

The RSPCA runs clinics and shelters for sick and abandoned animals, enforces the laws on animal treatment, and assists the passage of new legislation for the protection of animals (for example, the Companion Animal Bill, which is currently under consideration by the Victorian Parliament).

In their policy paper on farm animals, the RSPCA (Victoria) has put forward in detail the Society's position on what is and what is not acceptable practice. I agree with most of their recommendations, which take the rather difficult middle ground between those of groups that oppose the use of animals in any way and the supporters of intensive farming systems.

The RSPCA is concerned that some methods of intensive farming cause an unacceptable degree of stress to farm animals. It is opposed to systems that deprive animals of the opportunity to exercise and enjoy their natural environment.

People who keep animals on a small scale are unlikely to get into intensive systems. But they too should consider the need to allow the formation of social groups. It is important to keep animals with others of their own kind. A couple of goats or sheep or pigs is the minimum number. (Sheep actually prefer a minimum of five.) A cow will be happy with her calf, and often a horse will team up with stock of another species for company.

(Royal Society for the Prevention of Cruelty to Animals (Australia), 4 Lyell Street, Fyshwick, ACT, PO Box 462, Fyshwick, ACT 2609, phone (062) 80 5400. There are branches in every state.)

The Australian Federation for the Welfare of Animals

The AFWA seeks to put forward a humane and moderate approach to animal welfare. It is described as an 'independent national body of people who rely on animals and wish to put common sense back into animal welfare'.

The guiding principle is that the members of the Federation acknowledge that the welfare of human society is dependent on animals for food, clothing, health, education, recreation, and companionship, and that animal-related activities are legitimate provided a high standard of animal welfare is maintained.

Membership of the AFWA is open to any group interested in animals. The objectives of the Federation are to promote high standards of animal welfare, while raising awareness about the interdependence of humans and animals. The AFWA was formed to enter the public debate about animal welfare 'to redress the one-sidedness of the debate'. It claims that 'this debate has been dominated by groups whose aim is to abolish all forms of animal use by humans'.

The AFWA has published a code of ethics, to which members must adhere. The code recognises that:
- Persons using animals have an obligation to provide humane care and treatment to their animals.
- Animals feel pain and experience suffering.
- There is a need for all animal users to prevent unnecessary pain and discomfort to animals in their care and control.

All members of the AFWA must promote:
- Encouragement of the provision of adequate, trained supervision of all animals.
- Discouragement of cruelty, abuse and neglect of animals.
- Acceptance that they must provide for the needs of their animals as set out in state and national codes of practice.
- Adherence by members to those codes of practice that apply to their particular situation or, where necessary, development of a code of practice in which the humane care of animals is emphasised.

(Australian Federation for the Welfare of Animals, PO Box 908, Blacktown, NSW 2148, phone (02) 631 8022)

Keeping animals happy

Even on a small area it is possible to plan for contented animals. The likes and dislikes of each species are indications of what we should provide for them. Try to keep at least two of a kind of larger stock, and more of smaller ones. Keep the male to female ratio right for domestic harmony in poultry – the ratio varies according to the species. Poultry can be kept for egg-laying, meat production or grazing, and if you do not intend to breed from them, males will not be necessary.

Too many animals on a small area can be a disaster. This is one reason for the local council's limits on numbers. Too many animals can create a slum, with mud and dust, fighting and rats invading the area. A small paddock may soon be grazed bare, and erosion follow, with a heavy dose of parasites for the unfortunate stock.

However, there are ways of managing a small plot to carry a high stocking rate and improve the fertility. The aim is to create a pleasant and sustainable environment, which produces more energy than it takes to run. There are many good models to follow these days, such as the 'Permaculture' ideas of Bill Mollison and his colleagues in Tasmania.

Let us look at an example of an animal that can either create a desert or produce food from waste, depending on management. A couple of goats are introduced and they eat up the blackberries. So far so good. When they get into the garden, they are banished to a small paddock, eat it out, and look round for more. At this stage, in many backyard sagas, exit the goats.

Some of the solutions I have found to work for goats are:
- Tether them to a movable hut and move them frequently, bringing to them things they can't reach, such as branches. A shrub or two of goat-fodder, such as tagasaste, can be included in a planting programme.
- Give them a concrete yard in front of the shed that they sleep in. There they can get fresh air without being on the land. (This goes for pigs as well.) A few times a year, when the land is too wet or growth is nil, secure the goats in the yard to give the paddock a rest. They will, of course, need adequate shade, but will probably get it in the shed.

- Go foraging for weeds, branches and so on, and bring them back for your goats. If you live in the country, this should be no problem.
- Take them for walks, so they can do their own browsing. But keep them off the neighbours' favourite trees! With a proper lead and a harness, I have found goat-walking quite pleasant. It certainly teaches you what they like to eat.
- Tether them on the nature strip for short periods, checking frequently on shade and water. Tethering is a great idea if done properly, but never leave a tethered animal for long periods without checking on it, because many things can go wrong; the most common mishap is tangling up the rope until the goat nearly chokes.

Tethering is useful to keep animals from straying, but it needs proper arrangements and careful supervision. It is not suitable for pigs or birds.

Horses, donkeys and cattle need a leather head collar. Goats and sheep need a leather neck collar. A swivel is also needed, and at least six metres of rope or chain, fixed to a central point. Shade and water should be within reach. If you tether for long periods, let the animal have exercise. Don't tether young animals. They need to be able to run about.

Tethering by roadsides is quite often done, but anywhere potentially frightening is obviously not suitable. Also, the grass along roads with a lot of traffic will be polluted.

Where tethering is for grazing, the site should be changed every day; there should be plenty of grass, but no poisonous plants. Tethered animals should be inspected frequently, especially in hot weather.

Suburban goat-keepers can forage at the shops. Greengrocers, supermarkets and bakeries have stale food to give away or sell cheaply to foragers. Stick to uncooked vegetable waste, and there should be no disease problems.

These activities are cheap but time-consuming. They make good activities for children and families to do together, and they can be most relaxing after a day of earning a living somewhere else.

Keeping animals happy does mean regular attention to them: not only feeding and watering, but daily observation. This can become a chore. What about holidays and weekends away or even work-related trips? For some people, this is the sticking point, and it is sometimes the reason why families stop keeping animals after a year or two. It is the biggest single negative factor.

There are various ways to share the load of keeping animals, bearing in mind that the final responsibility is yours. Some solutions are:
- Cattery or kennels for pets.
- Neighbours with whom you can reciprocate.
- House-sitting could include tending animals, if you go away on a long trip.
- Farming out grazing animals.
- At certain times of the year and under certain conditions, grazing animals can be looked at once a day and need little else. This could be done by a competent neighbour.
- Responsible family members could give each other time off; for example, by following a roster for milking or weekend duties.

What about the neighbours? In country districts, most livestock will not raise an eyebrow among the locals, but in a suburban area, things could be more difficult. Neighbours may be alarmed if livestock is brought into a residential area. And even if you have checked the local laws and know that you are doing nothing illegal, it is still better not to antagonise people in the community in which you live.

Communication is usually the answer. Get to know the people around you and let them get to know you. Explain what you are doing, and try to get them interested in it. Make sure that they are not annoyed by the sight or sound of your stock and take great care that there is no smell for them to complain of. An old trick is to give manure to gardening neighbours, as this will make any animal odour more acceptable. Small gifts of produce from time to time will also sweeten the atmosphere. If you let neighbours meet your animals and take an interest in them, you may end up with a support group on your side, instead of criticism. The old magic of animal therapy could apply to potentially critical neighbours.

Naturally there has to be give and take. There is usually room for laying hens in a small backyard, but a rooster may make too much noise too early. There is no need to keep a rooster, because hens will lay happily without one. If you want a broody hen to sit on a clutch of eggs, you should be able to get some fertile eggs from someone who has a rooster.

Alternative medicine for animals

The health of animals in commercial farming has not improved much over the last thirty years, but the cost of medicines has risen faster than other production costs. We should question, therefore, whether the intensive systems of keeping livestock are healthy. Does overcrowding in itself affect the health of animals? Diseases that cause real problems are rarely seen on small farms, where we can care for the individual animal, making sure that each animal gets enough space, fresh food, water and shelter to keep it healthy.

Natural resistance to disease is the key to organic systems of animal health. Immunity is built up by young animals, who produce antibodies as part of their defence system. The other defence is the white blood cells, which fight infection in the healthy animal.

And what about vaccines? Should we vaccinate as a routine, as most people do (for example, against clostridial diseases in sheep)?

I used to think that vaccinations were acceptable because they stimulate the animal to produce its own antibodies. But organic scientists think that vaccines may be bad for the immune system as a whole, even though they are effective for a specific disease. The poor old sheep who are being hit with progressively more complicated mixes of bacterial vaccines are showing a shock reaction to being given up to sixteen different agents at once. The organic approach is to use vaccines only where the animals are really at risk: for example, when the disease is actually present on the farm.

Prophylactic drugs, now very common in farming, are avoided by organic farmers. That is, they don't dose the stock as a routine to suppress infection. There are several reasons for avoiding routine use of drugs:
- Regular drugs tend to lower natural resistance to disease.
- Some strains of bacteria have increased their resistance to drugs and have become more virulent as a result.
- Drugs can persist and get through to humans who eat the animal products, causing allergic reactions in some people and resistance to drugs in others.

One example of prophylactic drug use is dry cow therapy for mastitis in cows. This involves infusion of the udder with long-lasting penicillin, which will persist for several weeks while the cow is not milking. Most vets recommend that every cow is treated with a tube in each quarter after she is milked for the last time, before resting prior to calving, which starts the next lactation.

People who take their vet's advice and use dry cow therapy are not villains. They are trying to do the best they can to fight mastitis, a most painful disease for the cow, and one that costs farmers a lot of money in lost milk. I've used it myself on cows with a history of mastitis, and I hope it saved them some pain. Keeping strictly to organic principles may sound straightforward, but sometimes the choice is not an easy one.

Nursing

Vets have often told me that nursing animals is out of fashion. It went out with the introduction of the hypodermic needle to farms. People now expect a high-cost drug to effect an instant cure, but the old-fashioned way was different. It involved looking after the animal, giving it quiet surroundings and time to recover on its own.

Animal nursing takes time, it is worrying, and it can be heartbreaking when the creature dies in spite of your efforts. But it can be rewarding too. The procedure will vary according to the type of livestock and the disease involved, but it goes like this:

- Isolate the animal in a quiet shed, or outside in a small paddock handy to the house, with plenty of clean water (and bedding, if indoors).
- Feed very little, or not at all; molasses in the water for energy, perhaps.
- Keep an eye on it; talk to it quietly.
- Bathe wounds with herbal antiseptic, such as rosemary, and wrap them.
- Milk out infected udders frequently, sponge with hot and cold water, and massage the inflamed area. (This can apply to cows, sheep, goats, pigs or even rabbits with mastitis.)
- Dose with garlic for infections or parasites.
- Avoid sudden shocks, especially with horses.

Herbal medicine

It is open to everyone to experiment with herbal medicine. Herbs are easy to obtain and easy to grow. The treatments will usually do no harm, even if they don't work. There is a wealth of information about herbs in every library and bookshop. Two books I have found most useful are Mrs Grieve's *Modern Herbal* and Juliette de B. Levy's *Herbal Handbook*.

Plants contain several kinds of medicine. Some are astringent and contain tannin, which heals and cools; others have mucilage, which soothes internally and externally; some are bitter and tonic, some laxative or binding; and all plants are good sources of vitamins and minerals. You will learn with practice, but watch out for poisonous plants and be sure you identify them correctly.

Comfrey is one of my favourites for animal treatment, both internally and externally. It contains allantoin, which causes cells to grow quickly. It is most useful in an ointment, for example for a horse that has been cut on a wire fence. As well as being a high-protein fodder crop, comfrey helps to check scours in calves.

Garlic is very antiseptic, and is recommended for normalising blood pressure, both high and low, and is useful as a mild wormer.

Chickweed is a useful wound dressing (it cools inflammation); so are daisy leaves, and plantain leaves stop bleeding.

These are just a few examples from a vast range of useful plants.

Homoeopathic medicine

Homoeopathic medicines are gaining in popularity and are now being applied to animal health. Many vets and farmers in England, for example, now practise homoeopathy, but this is not a new fashion. The founder of modern homoeopathy was a German named Samuel Hahnemann (1755-1843), who referred back to the ancient Greek physician Hippocrates.

Homoeopathy has been defined as a therapeutic system in which diseases are treated with substances usually in extreme dilutions, which, when given to healthy individuals, produce the same symptoms as the disease they treat.

Substances are given in such a dilute form that it is difficult to believe they can have any effect at all. This is why traditional medics find homoeopathy hard to accept. But one advantage of administration in a highly diluted form is that there is no risk in trying them.

Homoeopathy is intended to treat the whole animal and to raise its level of resistance naturally, but the treatment depends on the 'character type' of the patient, so each treatment is individual. This seems to me to be a good approach to healing, but it means there is a lot to learn. Homoeopathy is more complex than herbal medicine.

Many homoeopathic remedies are preventative, such as dosing dry cows with minute dilutions of calcium and phosphorus and magnesium in order to safeguard them from milk fever at the beginning of the lactation, or putting 'nosodes' of mastitis bacteria in the drinking water of dairy cows.

Probiotics

On one occasion I used a probiotic treatment, prescribed by the vet, and it seemed to be successful in treating scours in calves. Probiotics are the opposite of antibiotics. They are 'good' bacteria, which restore the gut flora of a sick animal. Yoghurt is sometimes prescribed, and the tablets I was given for calves contained a culture of acid-producing bacteria.

Experiments are now being carried out to control salmonella in poultry by inoculating young birds with a culture of the bacteria found in the gut of healthy adult birds.

Disposal of farm animals

This issue must be faced. How will you dispose of animals that are sick and suffering, too old or surplus to requirements? A simple code of practice would be as follows:

Don't allow your animals to breed unless you can find the space to keep the offspring or can find homes for them. Sheep have only one or two lambs, but sows can have up to ten piglets.

If there are bad climatic conditions, such as severe drought, decide which animals to keep and dispose of the rest before all suffer.

If an animal is suffering and is not likely to recover, reach a decision quickly and have it humanely destroyed. Some people agonise for too long.

Animals should be humanely slaughtered by means of a mechanically operated instrument, or stunned before killing.

Animals not fit to travel should be destroyed on the farm.

Many farmers possess firearms and are usually willing to shoot a suffering animal for the owner. Some abattoirs provide an emergency slaughtering service. In the hands of an experienced person the firearm should be adequate for the job. For example, a pig or sheep can be killed with a rifle, revolver or shotgun (12 bore or 20 bore) with a shot not smaller than No. 5. The animal should be given some food and then shot behind the ear from a distance of 1 – 2 metres (or 3 – 5 metres if a rifle or revolver is used), so that the shot passes through the brain.

Slaughter of goat kids is often necessary on a farm if they are males. An overdose of barbiturates is often used, or a humane killer. If drugs are used, the meat should not be fed to other animals. A beef animal that has to be destroyed on the spot is usually disposed of by a licenced slaughterer using a captive bolt. Piglets that need to be killed can be dispatched quickly with a sharp blow to the back of the skull with a hammer. For hens, chickens and ducks, cervical dislocation is the most humane method, but it has to be learned. We were taught how to do this at agricultural college. Even if you do not intend to produce poultry for meat, it may be as well to learn how to kill them in case of an emergency.

The best way is to have an emergency procedure thought out, ready for use if necessary. It may be years before you have to use it, but one day it may save animal suffering. The plan may save you from having to kill an animal yourself if you would prefer not to do it.

Small carcasses, such as piglets that die at birth, should not be thrown on the compost heap but buried carefully in a place where they will not contaminate water. Large animals may be disposed of to plants that recycle them; for example, for blood-and-bone meal, or to kennels for dog food. Work out what the procedure is in your area.

If you have the facilities, and there is no risk of disease, a carcass could be cut up and used for feeding your dogs, if there is space in the freezer for it and you can face the task.

3

The Backyard Farmyard

This business of self-sufficiency

Can we really produce enough for our own needs? The killing and eating of an animal we have reared can be a big problem. Some people can do it, some can't. Those that can't often include professional farmers with years of experience, so it's nothing to be ashamed of. But we don't have to kill our stock. Vegetarians can keep animals to produce eggs, honey, milk, yoghurt, butter, cheese or fibre.

Meat production is easier if we get to know the breeding stock, but do not get too attached to the offspring. A rabbit or two from a large litter may well remain anonymous, and by the time it has grown into a sheep we may have distanced ourselves enough from a lamb.

I knew a small flock of sheep that had all been reared on the bottle as orphan lambs. They were all well known to the owner. Fortunately they were all female, so they went on from year to year as breeders. The last of the originals died of old age last year at the age of twenty or so. So if you rear an orphan lamb, try to find a female, because a male should be fattened for the freezer, and you will probably find it too difficult to accept your orphan as lamb chops.

Another problem with self-sufficiency is the treadmill, simply trying to do too much. Specialisation of labour developed quite early in human history. It enabled people to develop special skills, and introduced concepts like leisure and creativity. I enjoy growing food, but I also enjoy thinking, reading and writing, for which there would be no time if I tried to be totally self-sufficient.

The advantages of keeping our own livestock are obvious. We save money and gain much satisfaction. In the process we learn a great deal, enjoy the life-style and, best of all perhaps, our food is fresh, unprocessed, and free from additives. The humble egg just tastes a lot better fresh from the nest.

Help, advice, and training

Local laws have replaced by-laws in Australia, although some councils are still sorting out their local laws. In our area, a law that required a permit for moving stock on a road, even down the road for milking, was amended after protests.

The permit was really intended for people droving or moving stock long distances on foot.

Local councils have law enforcement officers, who see education as a major part of their job. They will give you useful local advice.

Enforcement of other regulations about keeping livestock is the province of your state department of agriculture. There are advisers in the department who can help with all areas of livestock husbandry – not just regulations – and they are backed by technical information from government research stations. Budgets for research have been cut heavily, but it does still go on.

One of the areas investigated by government scientists is the biological control of pests. Research is also being done on organic farming methods.

Agricultural colleges are becoming more accessible to people with small acreages. It would be wise to check your local college to find out what courses are offered. Some of the colleges offer weekend practical courses in skills like fencing, which are not easy to learn from books. Some agricultural and TAFE colleges offer distance-learning courses on subjects such as 'Trees on Farms', to quote one from our nearest college.

In Australia, learning is becoming more flexible as we move to competency-based training. Instead of having to undertake a two-year diploma course in farming, you can pick out one or two modules that are useful to you, and join a class just for the bits you need. Age is no barrier and classes are very mixed.

It costs money to take training courses, more so since the institutions have been asked to become more self-funding. Training dollars are money well spent if you research the course first.

In some areas the local SkillShare project will offer training in land-based skills. These are free to unemployed people, but possibly accessible to anyone for a fee. If you are unemployed, training in general will be cheaper for you, because there are usually concession rates for pensioners and people on benefits.

Increasingly, private training is available for people who want to learn to spin, to make cheese and to work with horses. At the time of writing the most popular topic for study seems to be permaculture. The concept was developed about fifteen years ago in Tasmania by Bill Mollison and David Holmgren, and it is all about bringing ideas for sustainable systems of farming and living together into an integrated whole. It has developed into an international movement, studied in 138 languages, entirely without government support. A recent issue of *Earth Garden* magazine invites us to spend a warm winter learning permaculture in the subtropics, experience a self-sufficiency weekend in north-east Victoria, or stay with Anglican sisters on a rural property to reflect on the meaning of life and learn crafts.

Neighbours can be a great source of help for new farmers. Neighbours are at hand, and they know the local conditions. Some of our older identities make dry comments about 'New Age' farmers, but you pay people a compliment when you ask for advice and, as a rule, they will be only too pleased to help. Listen and note everything, then make up your own mind.

Veterinary surgeons are expensive, but essential. I've worked with vets when writing training programmes and they all say the same thing: 'Why won't people use us as a source of advice rather than just as an emergency service?'

Ask your vet to pay an advisory visit, say once a quarter, and walk around with her or him. Write down all their pearls of wisdom – vets are often great talkers. Let her look at the condition of the stock, their feeding and housing, parasite control, everything. Prepare questions, so as to make the best use of expensive time. Pay up gracefully and learn from the visit. Remember, too, that when there is an emergency the vet will be more effective, because she will already know your animals and their conditions. If you have discussed your views on medication and your willingness to try nursing before resorting immediately to drugs, this may one day be important.

Of course, prevention is hard to prove. How can you tell if your talks with the vet have actually kept problems away? As an honest vet once said when I thanked him for curing a sick pig, it's often impossible to tell whether an animal would have recovered on its own or not.

Departments of conservation offer help and advice on all forms of land care. One way of getting in touch with the department may be through your local land care or farm trees group. These informal local groups of people get together to solve common problems, such as salinity or erosion.

If you join a local goat society, you will be invited to lectures, shows and demos, and you'll have access to goat experts. The same goes for horses. The local pony club is a very good place for children to learn, as well as for meeting horsy people. The local newspaper will give you news of activities, and your library will have the names and addresses of the people involved.

Why not try WWOOF for practical training? Willing Workers On Organic Farms is a worldwide network which keeps a register of families who will welcome people (the willing workers) to their holdings, to stay for a while and work with them at whatever jobs need doing. No money changes hands. Instead, there is a fair exchange of work for bed, board and useful experience. Some people WWOOF it all over the world!

For people new to farming, this would be a good way of learning to milk goats or look after bees, before you take on the responsibility of doing it yourself. It is also, to judge by the reports, a way of making new friends. For people already on the land, registering in the network might bring you helpers to finish the mudbrick house or gather in the potatoes. The usual length of stay is a weekend, but some people take their annual holidays this way.

If you want to become a WWOOF-er, write to Lionel Pollard, Director, WWOOF, Buchan, Victoria 3885. If you want to become a host, Lionel will send you a questionnaire about your property and the sort of people you would like to host.

Equipment

Get fences in place *before* the animals arrive! Shelter and food and water should be carefully worked out before a single hoof touches your yard or paddock. But the equipment can be accumulated gradually, as the need arises.

We have over 100 acres and we feed out the hay in winter with a four-wheel-drive vehicle and a trailer: you won't need a tractor. Even harrowing the grass can be done with a four-wheel-drive.

I used to wheel out the bales in a barrow, which is one piece of equipment I would hate to be without. A big barrow with a rubber tyre and no smelly engine is my idea of appropriate technology for small holdings. It is used for wood, harvesting potatoes, cleaning out the chookhouse and moving heavy loads from place to place. All backyarders need such a barrow.

When it comes to harvesting hay, you can either opt for low technology or hire a contractor. Either option is more efficient than owning the tractor, grass-cutter, rake and baler you would need to do it yourself. Small areas of hay can be cut with a scythe, turned with hay forks and raked up and stored in a shed, all with family labour. After all, this was the way our ancestors did it. The hay smells much sweeter when there are no diesel fumes around; and the work is hard, but much more satisfying than a workout at the gym. In cooler areas, tripods can be used to finish off the drying, as they used to do in the Swiss Alps. It works. A tripod is made with three pieces of wood, and the hay hangs on the frame so that it can catch the breeze.

Hand tools will come to you along the way. A kindly neighbour gave us a tool for rooting out thistles. His place was weedless, yet never saw chemicals: his only weeding was by hand. If you buy only what you need, and maybe borrow some tools until you know whether or not you need them, there will be no waste. Farm sales are good places for people who want to pick up second-hand tools and equipment. The following list is not comprehensive, but it will give you some idea of the sort of simple tools a backyarder needs.

For general use Spades, shovels, forks, hay-forks (two-pronged), rakes (wooden hay-rake and leaf-rake), hoes, axes, block-splitter, pincers, pliers, a screwdriver set, a socket set, tin-cutters, saws, hammers, assorted nails and staples.

For fencing Heavy hammer, crowbar, wire-cutters, heavy gloves, wire strainer, hammer, wire key for twisting wire, electric fence wire, and battery or solar fencer.

For monitoring Soil-testing kit, rain gauge, and weather vane.

For stock treatment Drenching gun, antiseptic (such as Stockholm tar), ear-tags and pliers, iodine, syringe and hypodermic needles, hoof-trimming knife, halters, and ropes.

For making dairy produce Stainless steel pans, ditto bowls, or hard plastic ones, fine muslin for straining, dairy thermometer if possible, colander, large spoon, large glass jar or small butter churn, cheese moulds or something that can be adapted for the purpose, clean wooden board for working butter, scotch hands or wooden bats for butter, starter culture.

For rearing poultry Incubator (or broody hen!), brooder, small run for baby chicks, drinker and feeders for chicks.

I don't like engines, so it pleases me to argue that the environment needs people who are willing to work by hand and walk round their land, just as some people manage to live without a car or a dishwasher. There are all kinds of attractive little ATVs (all-terrain vehicles), but they are not needed on a small acreage. They look like small-farm toys, but it's the big holding where they're needed to round up the cows or to ride round the fences.

You won't need a ride-on mower, either. The geese will keep the grass down nice and short, where grass is necessary as a fire precaution round the house.

Some people yoke the pony up to a little cart for working round the farm, although it's time-consuming compared with starting up an engine. You have to catch the horse, bring it in, gear it up and then supervise it while you are working. I've worked with horses a little in the past, and it is a good idea if you can find someone to help: an old-timer who knows how the collar should sit and where to get harness repaired. But few horses these days have been trained for pulling things, so it's likely to be a learning experience all round. I have seen my brother teach a horse to pull, and the first stage was to yoke it up to a log, to get it used to being followed.

Horses work without petrol, but they do need stoking up with oats. A horse can do a little light work on a diet of hay, but if you get your horse in more than once or twice a week either for riding or working, the diet should contain some high-energy oats.

There is a lot you can do with people power and hand tools when your land is small, and planning will cut down on the human energy needed. Grow perennial crops as far as you can, mulch instead of digging, and generally work with nature instead of fighting the environment. The most useful piece of equipment you have is your brain, and planning is a most valuable use of your time.

Managing the block

Good management consists of integrating the greatest variety of plant and animal life that is sustainable on a given piece of land. It consists of catering for the needs of all living things in the system: people, animals, plants and the soil. The good manager thinks a lot. Apart from the Department of Agriculture in your state, you could contact the following organisations for advice:

National Association for Sustainable Agriculture, PO Box 366, South Sydney, NSW 2000, phone (047) 51 4651.

Organic Farming and Gardening Society of Australia, GPO Box 2605, Melbourne, Victoria 3001, phone (03) 391 2470.

For the current addresses of breed societies, contact the Royal Agricultural Society in your state.

Any block with good soil cover, trees and water will look beautiful in time: shelter trees soften the hard lines of fences, climbing plants cover corrugated sheds and a stretch of water adds another dimension. Just keep planting and caring for plants until you find out what is useful and grows well. The livestock, sleek and contented, will complete the picture.

A properly integrated piece of land may look like this:

Grassy patch of lawn or paddock
If it is big enough, it should be subdivided to allow the animals to be rotated and give the grass a rest.

You may be able to graze a goat or two on it, and a couple of lambs or a cow and a calf. Rabbits and poultry can be kept in moveable runs, so you don't need to subdivide the land to rest the grass. Their rich manure will improve the soil. If you want to sweeten up the soil quickly for other stock, you can spread a little lime where they have been.

To protect the paddock from strong sun and driving wind and rain, and to provide some shade for the stock, plant trees to form a hedge. Choose your shelter belt from what grows well in your area: natives like tea-tree make a good hedge, as do fodder plants like tagasaste.

Orchard
Cultivate fruit and nut trees that suit your locality's soil, climate and shelter. The fruit needs sun to ripen, but a belt of mature fruit trees may shelter the vegetable plot which also needs the sun. In most parts of Australia you can grow vegies quite well in a garden which is shaded during the afternoon.

When the trees are well established and their lower branches out of reach of stock, you can do away with the paddock and practise two-tier farming, with geese or hens, pigs and eventually even sheep grazing between the trees. The trees provide shade for the stock, the animals keep the grass tidy, and when the fruit is ripe they eat the fallen fruit, which would otherwise rot and attract disease. Pigs can live in a rough wooden shelter in the orchard. (Our pigs used to get drunk on fermented apple juice from fallen apples, go to sleep, and snore loudly.) Beehives can be placed near the orchard, though not directly under the trees.

Pond or dam
Water beautifies the landscape: you can sit on your verandah and watch the sun set over the water, which reflects the last of the light and makes each twilight memorable. But a pond or a dam also provides fish for the table and a watery paradise for geese and ducks. Both nibble the water weeds, and ducks also eat the snails. During dry periods the water can be used to irrigate the vegetable garden and as a reservoir in case of fire.

Water also attracts wildlife and birds which help control insect pests. (Native plants also help attract birds.)

Garden
A garden for vegetables and flowers (and perhaps a propagation area with tree saplings and other nursery plants) provides food for people and animals. It is also beautiful, and the garden's soil will be enriched by the compost made with manure from the animals.

A well thought-out garden includes blocks of annuals, rotated each year: potatoes, corn, carrots, zucchini and so on. Such annuals, particularly if you let some flower, also attract predators of insect pests. Even during the southern winter our plot of annuals provides a variety of vegetables: in July there are Brussels sprouts, carrots, swedes, parsnips, silver beet, tree lettuce, celery and leeks!

Most of the vegetables we grow also provide food for rabbits, poultry or goats. If your land is big enough, you can grow potatoes to feed cattle (they eat them raw) or to feed pigs or poultry (they eat them steamed and mashed with bran). Or you can grow extra quantities of sweet corn for the hens: they love to help themselves to it if given half a chance.

Avoid digging, as much as possible. Digging destroys the soil structure and takes a lot of hard labour. Use perennials, our old friends which stay with us for years: Welsh onions, artichokes, asparagus, silver beet, rhubarb. There are many perennial vegetables. I used to have a perennial broccoli, and I would like to find another one to plant in our present vegie garden.

Apart from food and flowers, the rich garden soil will also produce high yields of weeds. Don't let them go to waste: rabbits love a weed salad, and so do goats. Any surplus should be returned to the soil as compost or green manure. If the cow produces too much milk for your needs, feed it to the chooks (they love it) or to the pigs or lambs. And apart from feeding the animals, surplus food can probably be bartered locally: just about every area has a lively bartering system. (We recently exchanged eggs for berries, onions and tomatoes.)

If the thought of all this teeming life scares you a little, start simply. Grow some fruit trees and get a few geese. They're hardy birds and they provide eggs, meat and feathers as well as being good lawn mowers and keeping intruders out. You may then add a dam or pond. The geese will love it, and you can grow some fish as well. Then you could perhaps acquire a rabbit hutch. If part of the hutch floor is made of wire mesh, to allow the droppings to fall into a pit beneath it, you can keep worms in the pit to compost the manure for the garden. You'll then need a vegie patch for the compost, and perhaps some chooks to scratch about in it in autumn, cleaning up the insects and getting their reward with ripe corn cobs. May be you will end up with a goat or sheep or cow as well.

Whatever you do, remember that the key to a properly integrated system is to promote harmony and balance as well as variety, and to return as many nutrients to the soil as possible. That way you'll have a very pleasant environment which will grow more beautiful every year, and you should still have time to sit on the verandah and watch the setting sun.

Building up fertility

Peasant agriculture was successful because animals were part of the organic cycle. We can improve our land in the same way. To make really good compost, you will need something like poultry manure to mix in with the vegetable waste. With soil improvement, production will increase.

Much is said these days about the fragile Australian soils and how they do not produce food as the soils of Europe do. Apart from the comparatively high rainfall in Europe, the difference is really in the replenishment of Old World soils over centuries. My land was cleared only about forty years ago, so what it needs is humus and more humus to make it productive, combined with good water conservation, so it can be kept moist in the dry season.

It is hard to improve a large area of soil quickly. Start small with a little garden, and you will be surprised how fast the household and garden waste will break down into plant food, especially in warm weather. In colder climates the compost heap can be covered with black polythene to warm it up. We have a plastic compost bin and several heaps of grass cuttings, weeds and kitchen waste. We intend to build retaining walls with sleepers to contain the heaps. Everything organic and biodegradable is added to the compost.

Compost is the friable, rich, dark material that results from the bacterial breakdown of wastes. It adds plant nutrients to the soil, and it provided humus, the material that improves the soil structure with air pockets and helps it to hold water. Mulches and compost on the soil surface cut down watering, because they protect the soil from drying out, which it does very quickly in sunny or windy weather. Fresh lawn clippings make a good mulch.

The ancient system of composting and manuring works well on a small scale. It is sometimes tried on a field scale by organic farmers with mixed farms. The farmers of Europe used to spread the land in spring with manure accumulated from the winter housing of livestock, but in Australia only the pig and poultry industries produce manure in quantity. However, the value of poultry manure is being recognised, and you can now buy a packaged variety (such as Dynamic Lifter) for use on a farm. There are other ways of adding fertility to the soil, such as green manures and seaweed, but animal manures are such a great help to soil improvement that this is a compelling reason for including livestock in your backyard ecosystem.

The main reasons why organic farmers avoid chemical fertilisers are:
- Chemicals do not improve soil structure (a particularly important point with Australian soils).
- Chemicals can 'lock-up' soil trace elements by forming insoluble compounds; plant food has to dissolve in water before it can be taken up by the roots of the plant. Lack of trace elements leads to deficiency diseases in both plants and animals.
- Chemicals are expensive
- Chemicals are derived from oil or from mining (that is, from non-renewable resources), whereas organic fertilisers are part of the carbon cycle, renewable through plants, which recycle energy from the sun and elements from the air.
- Chemicals can be toxic to soil organisms, which are necessary to a healthy soil. For example, nitrogen fertilisers kill the bacteria living in nodules on the roots of clover – nodules that fix nitrogen from the air and make it available in the soil. In Australia and New Zealand, clover is a most important source of soil nitrogen, which helps the grass to grow.

Conventional farming scientists argue quite convincingly that a high world population can be fed only by conventional methods and that these include the use of chemical fertilisers and pesticides. They say that a farmer cannot make a living on the lower yields obtained by organic methods. The growing number of organic food producers are obtaining a higher price for their products, but in Australia this represents only a small fraction of total food production. A *Which?* magazine survey in May 1992 showed that one-third of consumers in the United Kingdom now make regular purchases of organic food, and two out of five of those not already buying would like to do so if prices were reduced. Produce-packers reported for the same year that sales of organic vegetables were up by 40 per cent on the previous year (*Soil Association Journal*, October 1992).

Public taste is inclining to organic produce for several reasons, and these are also good arguments in favour of running your backyard on organic lines:
- The taste of organic produce tends to be better.
- Health: no poison sprays in the air or toxic residues on food.
- Support of environmentally friendly farming.

Organic produce is increasingly labelled and certified as such to guard against fraud. Backyarders who wish to produce for sale will need to research their local market and find out how to register their produce as organically grown.

Fodder crops

Permaculture is a growing system whereby crops and animals are integrated in small areas. It is well worth your while to investigate how permaculture principles can be put into effect in your area. Permaculture principles demand you grow your animals' fodder yourself, so the following section sets out a few ideas for growing food (apart from grass) to feed your animals and fowls.

Tagasaste (Chamaecyticus palmensis)
In the Canary Islands, small farmers grow this tree as a windbreak, and harvest it for cattle food. Although the tagasaste has been known in Australia for over a hundred years, there has been a recent increase in interest in this plant.

Tagasaste seems to me to be an ideal plant for an integrated system on small acres. It is a legume (like wattles), which means that it is capable of fixing nitrogen from the air in nodules on its roots, and this eventually becomes plant protein. It is quite a pretty tree, producing small, white flowers, which are visited by bees. Leaves and branches may be used for stock feed. The seed pods contain seeds, which make a high-protein poultry food. It is also called 'the living haystack' because, like hay, it can be kept for use in a drought.

We planted some very small Tagasaste trees last year, and they are now growing well in the sheltered sites. They did not grow in poor draining areas. They thrived in an unusually wet summer, but I am told that they grow just as well in Western Australia, at Tamin, where the rainfall is only 350 millimetres per year. Tagasaste will not withstand heavy grazing, but you can graze them lightly once they are 50 centimetres high. Regular grazing will minimise the woody growth, which animals won't eat. Even if you reserve the trees for times of drought, it is probably best to harvest them at least once a year, maybe by pruning to the height your animals can reach, in order to keep the growth young and palatable. Once your 'tags' are mature, you'll be able to cut or graze them in spring and again in autumn, and they should last for fifty years.

Propagation   Seed germination depends on whether you have the right bacteria in your soil. If inoculation is needed, the inoculum can be obtained from Root Nodule Pty Ltd, 84 Rawson Road, Woy Woy, NSW 2256. The seed has a waxy coating, which keeps out moisture, and this has to be broken before germination, either by immersing the seeds in boiling water and leaving them until the water cools, or by scarifying the outer coat.

Our trees were not grown from seed, but transplanted as seedlings. I dug them up from under some mature tagasaste trees, where they had germinated from fallen seed. If you know anyone with a few trees, you could do the same, or you could harvest the seed pods from the trees and try germination.

Native fodder trees

I have watched our cattle nibble round the bushy edges of their paddocks, so many native trees are good cattle fodder. In a drought, the greener and leafier species would be the most use. Gum-leaves are toxic, except for koalas, but there are plenty of wattles and casuarinas that will give the stock a change of diet, as well as supplying shade and shelter. In the drier areas, mulga (*Acacia aneura*) is grown for fodder.

Goats will browse much more than cattle, and they enjoy a wide range of native species, as well as weeds. Here is a list of a few natives of south-east Australia, which could form a shelter-belt while providing browse for ruminants:

Botanical name	**Common name**
Acacia melanoxylon	Blackwood
Acacia salicina	Native willow
Casuarina stricta	Drooping she oak
Cassina acculeata	Dogwood
Exocarpus cupressiformis	Cherry ballart
Leptospermum lanigerum	Woolly tea tree
Melaleuca ericifolia	Swamp paperbark
Pommaderris aspera	Hazel pommaderis
Pittosporum bicolor	Banyalla

Paulownia trees

One day last winter, a colleague gave me a few dead sticks in a hessian bag and implored me to plant them. So I did, eventually, after they had languished for weeks in the shed. And in spring they sprang to life, growing a foot or more, with huge green leaves. Paulownias are incredible survivors.

The Chinese have been using these trees for timber for years, harvesting them at ten-year intervals, for shelter, fodder and honey. In a competition organised for growers by suppliers called Farm Fodder Trees Australia, the winning tree, which grew in Queensland, increased over three-and-a-half centimetres a day for five-and-a-half months. For a fodder and shelter belt in the tropics, this tree seems to be a winner. It is also quite tolerant of adverse conditions: one I knew was stolen three times and survived!

Willows
Willows are out of fashion now, and are being replaced along the rivers by native trees. But willows can be very useful, to prevent erosion and as fodder for stock. Willow leaves are palatable and will feed any grazing stock in a drought, and goats at any time. They are fast growing, so you can let the goats loose on them at least twice a year. They are a good tree for coppicing; that is, you cut down the tree for firewood and leave the stump in the ground, where it grows side shoots and produces three trees where one grew before.

Herbaceous plants: Comfrey (Symphytum officinale)
This is an undervalued plant, which is easy to grow, perhaps too easy, because it is difficult to eradicate once you have it on your land. The leaves are large and rather hairy, and the plant grows to about a metre in the summer, dying down in the winter in the cooler areas.

I have grown comfrey for years, and I love to hear the bees humming among the blue flowers on a warm day. It is a good stock feed with high protein, once the animals get used to it, and it puts a shine on the coats of cattle and horses. There is a folk tale, which may be true, that comfrey was grown at the inns where stage coaches stopped, and that the horses were fed on comfrey at each stage.

Comfrey has acquired a bad reputation, because it is said to contain a substance that damages the liver. It is certainly a healing herb; and I make comfrey ointment every summer for use on people and animals. I use the leaves as a high potash mulch for potatoes and in water as a liquid fertiliser for tomatoes. Where the soil is fertile and there is plenty of water, comfrey gives a high yield, and the crude protein content is over twenty per cent. It is not a drought plant, because of its need for water, but it will give valuable fodder from a small space.

Comfrey is a hybrid form and does not set seed, so propagation is by root division. Small pieces of root will grow quickly in moist and warm conditions.

Milk handling and milk products

See also the sections on cows, goats and sheep in Chapter 4.

Starter culture for making dairy products is available by mail from the Rural Store, Lowdens Road, Kilmore West, Vic. 3754.

Milking the animals

Good hygiene is most important in dairy work, especially with goat's milk, because milk is a perfect breeding-ground for bacteria. The 'goaty' taint that puts so many people off the milk is easily picked up if the surroundings are not clean. It is best to have a clean place set aside for the milking.

I assume you will be hand-milking. Milking machines are very efficient for use on commercial farms, but the chore of washing them is greater than the chore of milking one or two cows, goats or sheep by hand. Most backyarders still milk by hand. Some people find hand-milking a strain on arm and hand muscles, and there are small portable milking machines with either petrol or electric motors.

Goat-keepers find it useful to have a raised platform, so that the goat is at a convenient height. Goats love to skip up onto a platform, particularly if it holds a feed bucket at one end. Designs vary, the platform may have a seat for the milker, a stand for the feed bucket, and a head-bail to hold the goat round the neck while she is eating. If you do not make a platform, keep a clean area of the shed for milking, away from the bedded area. Goats' bedding is dusty, with the potential for adding millions of bacteria to the milk.

Cows can sometimes be milked in the paddock, but anchor the animal before you start, either with a halter tied to a post, or perhaps a head-bail, which will be easy to fasten round the neck if there is some tasty food at the other side.

Sheep are docile creatures, on the whole, and are likely to be easier to handle for some people than goats. They browse a little, but are not as demanding as goats, nor as destructive, since they are easier to fence in. Sheep are also more versatile than goats or cows since, apart from a crop of wool, a ewe should produce at least one lamb a year and can be milked as well. If you are interested in spinning, it should be possible to tame a ewe sufficiently to milk her. A good ewe should produce more milk than one lamb needs, so the surplus could be used for home dairying.

Sheep were probably milked before cows were, and there are still a lot of milking sheep in the world. In Europe, certain breeds such as the Friesland have been selected for milk production, but many sheep milkers just use good young ewes of whatever breed is handy. In New Zealand experiments with sheep not selectively bred for dairying showed them to be capable of good milk production.

The problem with any milking animal is that its young need milk too. Sometimes the lambs are taken from the ewes at birth and reared on the bottle or lamb bar, at first with colostrum from the dam but later with a commercial lamb milk replacer, in much the same way as calves are reared on a dairy farm.

Alternatively, lambs can be kept on the ewe for five or six weeks and then weaned so that the ewe can be milked. New Zealand figures show that about 150 L can be obtained from one ewe during a milking period lasting about six weeks.

Sheep's milk is twice as rich as that of cows or goats. The fat content is usually over 6% and the protein content about 5.6%. Ewes' milk has small fat globules, so it freezes well and makes good cheese. Like goats' milk it contains no carotene, so its products are pure white. The milk's richness makes it attractive for 'gourmet' cheeses like fetta, and our cultural diversity means there is a growing demand for such cheeses of Mediterranean origin.

Lambs grow very quickly and thus need a high-energy food. To produce such a food sheep need a high-energy diet, so milking sheep should have some concentrate feed as well as good quality grass.

Milking animals like routine, and quickly adjust to it, so try to follow the same sequence at each milking. Twelve-hour intervals for milking are the best, but most people find this too much of a chore and strike a compromise.

The Backyard Farmyard 29

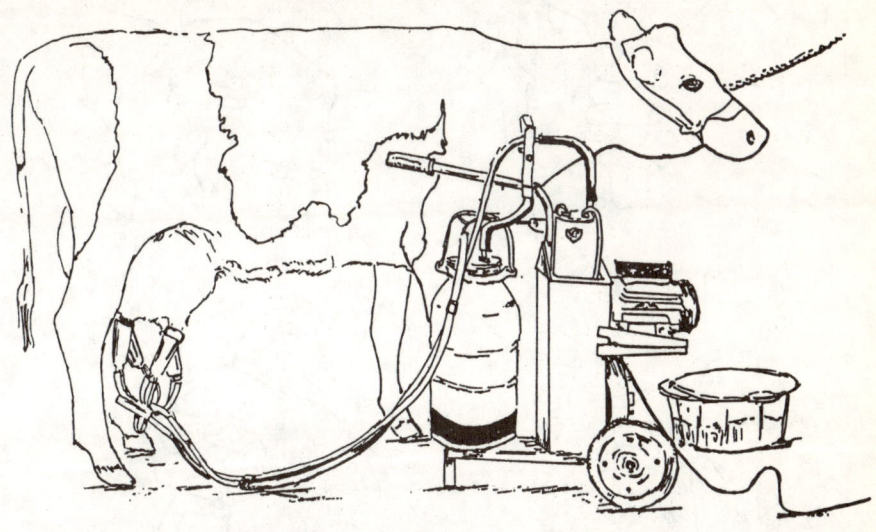

A portable milking machine, suitable for a house cow or small herd.

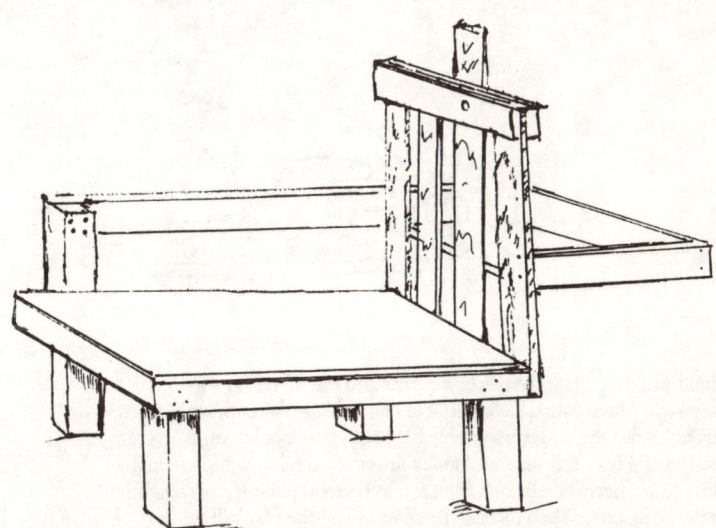

This is an American version of a goat milking platform. A young pregnant goat should be lured onto the platform with food and handled gently while she eats. This will become part of the routine, so that when she comes into milk, she has no fear of being milked.

This platform is at waist height, to allow the milker to stand when milking. Some designs have a seat for the milker.

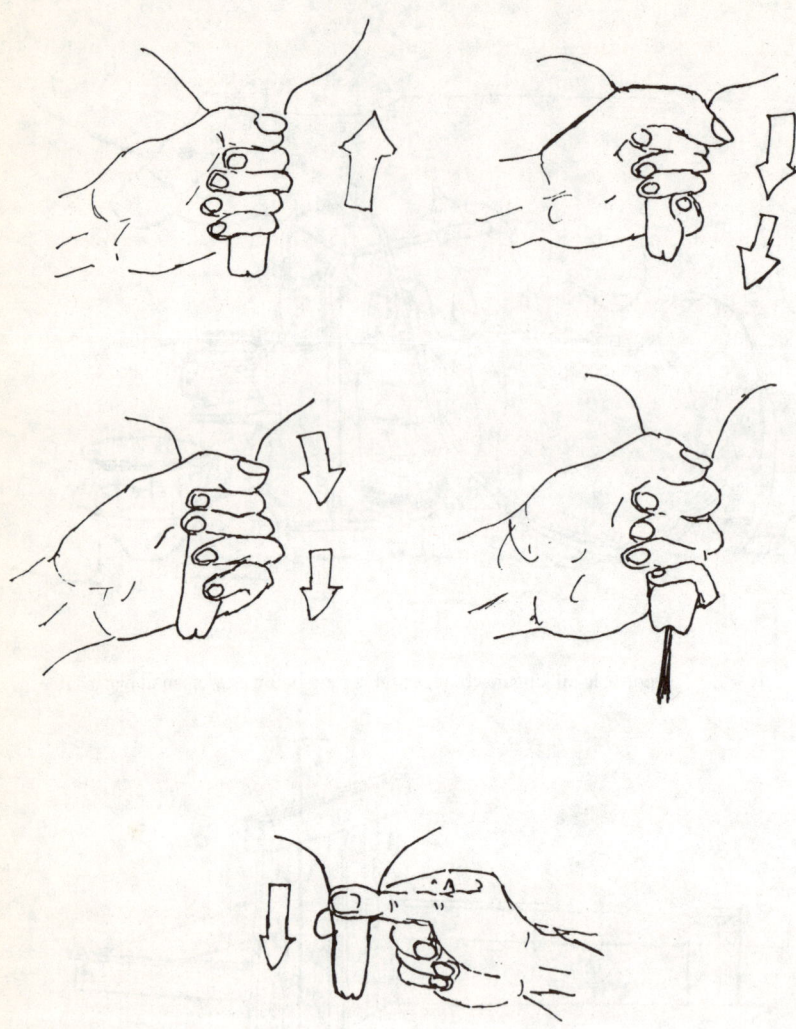

Hand milking Handle the cow gently and talk to her as you approach. Make yourself comfortable beside her on a stool, with a bucket under the udder, preferably held between your knees. Squeeze the top of a teat in thumb and index finger of each hand and gradually close your fingers one by one from top to bottom, pulling down slightly at the same time. Then release the tension and the teat will move upwards slightly. You will feel it fill up again with milk and then you can repeat the gentle pull and squeeze.

Stripping at the end can be done with the index finger and thumb. When you have finished the udder will feel slack. Don't worry about the very last few drops: milk production is a continuous process. If teats are sore there may be trouble. Udder cream will keep them supple and soft, or you can use lard as the old farmers used to do.

Learn to milk before you get your own cow, goat or sheep. Be gentle, but firm, when milking. When the animal is settled – probably eating concentrate – rub her udder and then, holding a teat in each hand, close your fingers one after the other down the tube, bringing the milk down with them.

The cow has four teats, so you need to change over after a few minutes, to give the other side a go. The four quarters holding the milk are almost separate, so it is possible to milk out one while the others are full, but even milking is the best. There may be a few moments' delay before the milk comes. It is 'let down' in response to the hormone oxytocin, stimulated by the butting of the calf or kid looking for food. In a dairy, the stimulation may be the routine: jumping onto the platform, being given feed. It is important to keep milking while the hormone is flowing: goats milk quite quickly as a rule and cows don't take more than a few minutes.

Adrenalin is produced when the animal has a sudden fright, and this inhibits oxytocin. So the aim of every milker is to keep things quiet and easy for the animal. This is why visitors can be a problem at milking time!

Cool the milk immediately. If you put it in the fridge, stir it while cooling. Old-fashioned milk-coolers were quite effective, but may not be available. The simplest way was to stand the milk container in the creek.

Quick cooling is important for two reasons: to slow down the growth of bacteria, and to prevent enzymes from breaking down the milk fat, which is what gives milk a strong flavour. After use, scrub the milking buckets and any utensils with hot water and washing soda and sterilise with boiling water. This is the traditional way, without the use of chemical sterilisers. This type of cleaning is best for three reasons: it is cheap; it involves less pollution; and there's no chance of flavouring the milk with a chemical taint.

Milk products

Milk products involve a lot of work, but they are satisfying to make. They enable you to use large quantities of milk, and to store the products for future use.

Specialised dairy equipment is available from self-sufficiency suppliers, but for a start you can, in the main, make do with household utensils. Use stainless-steel vessels where possible, because they are easiest to sterilise. A dairy thermometer is a good investment. The dairymaids of the past used to test the temperature of the milk with their elbows, but it is hard to make cheese without a good thermometer, although you can manage without one for cream and butter, and possibly for yoghurt.

The flavour of milk products comes from the development of acidity by bacteria in the milk, which convert lactose, the milk sugar, to lactic acid. We can either make a rather variable dairy product in the historic way, or we can take advantage of dairy technology and use a culture of the right bacteria for the product after heating the milk to kill off the resident bacteria. This is just like using wine yeast, as opposed to letting the grapes ferment with the wild yeasts present on the grape skin.

Yoghurt

This is one of the easiest products to make, and goat's milk is particularly suited for it. You can practise making yoghurt before you get your cow or goat. Yoghurt keeps longer than fresh milk. Yoghurt is milk, soured by acid-producing bacteria, including *Lactobacillus bulgaricus*, and this helps our digestive system, which is why yoghurt is classed as a health food. You can buy powdered culture to make yoghurt, but the easiest way is to use good quality natural yoghurt as your 'starter'. Skim-milk powder is often added to give a thicker yoghurt. The bacteria need warmth to make them multiply, so the milk is heated and then kept at about blood heat in a vacuum flask or yoghurt maker.

Sterilise all the equipment with boiling water. Add 25 grams skim-milk powder to 1 litre of milk and mix it. Heat the mixture to 85°C and hold it there for several minutes. This is a form of pasteurisation, to kill off the existing bacteria in the milk. Cool to about 37°C, then stir in about 2 tablespoons of the yoghurt culture. (If you put it in the hot milk, it will be killed.)

Leave overnight to set in a warm place, or in a vacuum flask and when it is set, refrigerate the yoghurt. Add fruit or nuts for variety when set.

Cream

Cream is milk fat, the lightest ingredient, which rises to the top when the milk is left to stand for some time. Because the fat globules in goat's milk are small, the cream takes about thirty-six hours to rise. Sterilise a shallow bowl, pour in the milk, and leave the bowl in the fridge with a cover over it. Try not to move the bowl until the cream has risen. Scoop the cream off the milk with a thin saucer, a sharp spoon, or a proper milk skimmer (a metal disc with holes in it). For more than one or two goats it is easier to use an electric separator.

Butter

The gentle Jersey cow is the one to keep if you like lots of golden-yellow butter. Butter is made from cream that has 'ripened' or been allowed to develop flavour. The cream from several milkings is mixed and kept for a few days before butter is made. In this case, hygiene is particularly important, because you are relying on the bacteria present in the milk to produce a pleasant flavour.

The flavour will be fine if the resident bacteria are good acid producers, but if bugs from the manure or bedding have invaded the milk, they will be coliform bacteria, which break down protein to produce unpleasant tastes and smells. In order to be sure of a good butter, you can follow the same routine as for yoghurt: pasteurise the milk to kill off the natural flora, and then introduce some of the right bacteria. There are butter-starter cultures as well as those for cheese and yoghurt.

The dead-white colour of goat's butter may not be appetising, but you can add marigold petals or the vegetable dye annatto to improve its colour.

Small glass churns with paddles inside are now available, and you can also make butter in an electric mixer or by shaking cream from end to end in a large coffee jar. I have made butter in all sorts of churns, but you can't beat the coffee jar for simplicity if the quantities are small.

Cream for churning should be cool, but not too cold: about 13°C will be the average, cooler if the room temperature is high. It should be thin enough to run slowly off the back of a spoon. Very thick cream will not move enough to churn properly, because the fat globules have to collide in order to stick together.

Don't fill the jar or churn more than half full, and start off slowly, rotating the paddle or shaking the cream from one end of the jar to the other. Increase the speed as you go. It usually takes up to fifteen minutes to make butter, but it can happen more quickly. At intervals, take off the lid to ventilate and allow the gases to escape. You may be surprised at how much acid-smelling gas will be released.

Gradually, as you churn, grains of butter should appear, surrounded by watery buttermilk. Sometimes you miss the grain stage and get one large blob of butter, but blobs the size of a grain of wheat are the desirable size, because they can be washed easily.

If nothing happens after about fifteen minutes of effort, you may have what the dairymaids used to call 'sleepy cream'. Milk, being an organic substance, varies, and this is why dairying processes are variable, just like wine and bread making. To wake up sleepy cream, add a spoonful or two of boiling water and carry on churning.

Drain off the buttermilk and wash the butter with cold water a few degrees colder than the churning temperature. If you like salty butter, stand it in brine (salt water) for a few minutes at this stage.

Traditionally, butter is worked with 'Scotch hands', paddles made of smooth wood with fluted faces. 'Working' means squeezing gently to get rid of surplus moisture and to make the grains stick together into a solid. Do not smear the butter or be too rough, because that will make it oily. Keep it cool, shape it into pats, and wrap it in greaseproof paper. Butter will keep for weeks in the fridge and it also freezes well.

Cheese

There are many varieties of cheese, and it can become a fascinating hobby to make sheep and goat's milk cheeses. Once again, the old-fashioned recipes relied on the milk bacteria, but modern cheesemakers tend to heat up the milk to 70°C for thirty seconds to kill off the natural flora, after which a starter culture of acid-producing bacteria can be added. You can buy cheese-starter culture in powder form and reactivate it or, at a pinch, you can use live yoghurt. (Do not boil the milk, because this may change the protein and spoil the cheese.)

Having added starter, the basic recipe for all cheese is simple. Milk is kept warm and made to form a curd, usually by the addition of rennet or a vegetable enzyme. The curd is cut into pieces and stirred in the whey. Then the whey is run off and the pieces of curd stick together to form a solid mat, which may be further drained and heated before pressing.

Soft cheese is just the curd stage, with little cooking and no pressure. It is only slightly acid and has a short life. Hard cheese keeps a long time and is more acid. The bacteria used in cheese-starter are chosen to give the cheese the right texture and flavour. You can buy dried starter culture, pasteurise your milk, and make cheese to a precise recipe.

Acid curd cheese

The simplest cheese is made by allowing the milk to sour and then hanging the curd up in a muslin bag to drain the whey off. This is the way we always dealt with sour milk when I was a child, and it can be used for any unpasteurised milk that still has its complement of acid-producing bacteria.

Stand the milk in a warm place and add a little lemon juice to help to make a firm curd. When it has soured and thickened, pour it carefully into a piece of sterilised muslin and hang it over a bucket. The whey will drip out slowly; drainage depends on warmth, and is quicker at a higher temperature. After a day or two, you can scrape the curd off the muslin and mix it with chives, other herbs, or a little salt. Here you have simple, traditional goat's milk cheese!

The recipes for cheeses made with rennet are more complicated. Soft cheeses are left to drain naturally, while hard cheeses are put under pressure to force out the whey, which makes the cheese harder and easier to store for long periods. Blue cheese is made by adding mould spores of the right variety to the curd, and then inserting air-holes in the cheese as it ripens.

Fibre from goats, sheep and rabbits

See also the sections on these animals in Chapter 4.

Angora goats

The quality of the stock determines whether you get good fibre. Pure-bred goats produce much more good-quality fibre than cross-bred ones. Two crops of mohair a year are usual, in early spring and early autumn. A pure-bred doe should give you 4–5 kg of mohair a year, and a kid will produce half this amount in its first year.

Cashmere goats

The fibre is short and fine, ideal for luxury knitwear. This undercoat is found on goats that do not produce mohair; and it grows as a winter protection when the day length shortens. There is a coarse outer coat, which is present all the year round, and which protects the cashmere underneath.

Cashmere was traditionally produced in places like China, Mongolia and India. There it is a third crop, after meat and milk, from the herds of goats, and it is combed out from the coat. It is one of the finest and most expensive fibres in the world. For the last ten years or so, feral goats have been farmed for cashmere in Australia, and here they are shorn rather than combed, which means that the fleece has more lustre than when it is combed out as dead hair. The aim is to shear in late winter, just before the coat would be shed.

Cashmere production involves processing machinery to separate the fine hair from the coarse, so cashmere has not been a backyard product, but one for farmers seeking an alternative enterprise. However, some people hand-comb their goats and sell the combed fleece to a craft shop.

Sheep

A whole fleece contains a lot of wool and will take a long time to spin. The best way to learn, I found, was to pick up the odds and ends of wool that sheep leave about on bushes and fences. A medium length wool, which is not as fine as the Merino, is the easiest to spin.

Take up a bit of wool, pull it out and twist it in your hands. The result is a strong strand, as our ancestors discovered thousands of years ago. The first spinners used a spindle and so did I, until someone lent me a spinning wheel.

A simple spindle can be made by sticking a knitting needle through half a potato. This is the size and shape you need, although my spindle was made of wood. Tease the wool out to begin with, pulling the fibres apart with your fingers. You will perhaps feel small twigs and seeds in the wool and these should be taken out. The little knots of wool are separated, and suddenly you have a fluffy mass which seems much more wool than you started with.

Ready-made yarn is threaded onto the spindle to start off with, and then the teased wool is gathered into the yarn and wrapped round it. The spindle is given a twirl, the yarn is pinched out and pulled and then let go, and the spindle spins the fleece out until you have a rather uneven yarn. When the yarn reaches the floor, it is wound round the spindle and then you spin again: a slow process, but people walked behind the flock as the sheep grazed, spinning as they watched the sheep. It was not until the fourteenth century in continental Europe that people sat down to spin at a wheel.

Most keen spinners now use a wheel and you can buy one in a pack, ready to assemble, or you can make one yourself (see the *Weekly Times Handbook* for directions). Spinners 'card' the wool with wooden bats to tease it out, so you will need carders as well.

Angora rabbits

This is a long-haired breed of rabbit with large, floppy ears. It is one of the oldest breeds of rabbit, probably kept for its wool by the Romans. It appeared on Roman coins in the time of Hadrian. Some people think that, like the Angora goat, it came from Angora in Turkey.

Angora rabbits are very far removed from their wild ancestors and would probably not survive in the wild. They would not be able to run fast enough to avoid danger, and would probably get their long hair tangled in the undergrowth.

Angoras have been kept in France for hundreds of years, the wool being mixed with sheeps' wool and made into a yarn for expensive knitwear. French, German, and English Angoras have developed along different lines. The ones I know best are the fine haired English Angoras, which yield less than the European types. Although most Angora rabbits are white, there are several different colours in the breed. The wool grows to about 15 cm long if it is left without clipping, as it is by the show enthusiasts. Rabbits with long wool are difficult to keep clean, but ordinary working Angoras, called 'woolers', are shorn of their coats about four times a year. The feathery plumes on feet and ears, called 'furnishings' by the breeders, give the rabbits a very fluffy look.

The clip
Angoras can be clipped or plucked. In America plucking seems to be popular but in Britain they are usually clipped with sharp scissors or electric clippers. Clipping is quicker and takes about half an hour. You can expect perhaps 250 g of wool at each shearing, four times a year, when the coat gets to about 8 cm long. The rabbit is usually laid on its back with hind and forefeet tied, but it seems that they get used to shearing, which is done very gently, and do not object. When the rabbit is sheared it is given a medical inspection as well, and the nails are clipped if necessary. The ears in particular are checked for mite or canker. Of course a newly shorn rabbit can catch a chill, so they are given extra feed, and a nest box for shelter in cold conditions. Angora wool is very popular with hand spinners when blended with other fibres, but is difficult to spin on its own. It is not so greasy as sheep's wool, partly because the rabbit doesn't produce very much grease from its skin. The Angora wool has an electrostatic discharge which stops dirt from sticking to it as it does to sheep's wool. The fibre is hollow and light in weight. The diameter is 11 to 12 micron. There are three grades of Angora fibre when shorn:
- clean, felt-free, more than 8 cm,
- clean, felted, under 8 cm, and
- dirty.

Good rabbits which have been managed well and kept clean average about 75 per cent Grade 1 fibre. The quality is better in cool climates, but it may be hard to keep the fibre clean enough for high grades on a natural system. Angoras are normally kept in separate pens.

Using the fibre
My spinning experience is limited to sheep's wool either on a spindle or a spinning wheel. Angora wool makes me cough, so I haven't tried to spin it – I can't wear it, and you may have the same problem. People allergic to cats sometimes find Angora wool unpleasant. Those who do spin Angora fibre say that it takes experience.

It needs very little tension and a very light touch. It is so fine and slippery that it is probably best mixed with sheep's wool for a garment and of course, if you use another fibre you'll be able to make a garment sooner. One way to mix them is to spin a single thread of yarn with Angora and do the same with sheeps' wool and then twist them together to make a 2-ply yarn. Blending Angora with other fibres makes it last longer and wool gives it more elasticity. You can use natural or chemical dyes with the fibre and some interesting results are possible with mixed yarns which take the dye differently.

Meat production

Calves, fish, lambs, goat kids, pigs, poultry and rabbits can all be bred for meat production. See also the sections on goats, pigs and rabbits in Chapter 4.

In country areas many butchers provide a service for people who wish to have their stock slaughtered, cut up and returned to them. Home freezers make it possible to store meat for months, but fresh meat is still appreciably better. Backyarders who want to produce meat for their own consumption might find fresh poultry and rabbit meat or fish most convenient because it can be prepared when needed. Before the days of refrigeration there was a season for everything, something that's easily forgotten nowadays:

Spring	Summer	Autumn	Winter
Rabbit	Lamb	Pork	Turkey
Lamb	Chicken	Beef	Rabbit
	Goose	Pigeon	
	Guinea fowl	Bacon and ham	

Home slaughtering of farm stock is legal providing the meat is used only for home consumption. Killing poultry and rabbits is a skill that can be learned from an experienced person, but the larger animals need more expertise.

An on-farm butcher's service is not cheap, but it is excellent and I think the idea is spreading. The butcher comes at the agreed time and kills the animal humanely without removing it from its accustomed surroundings: the best way for all concerned. After butchering, the meat is chilled on your property in the butcher's mobile coolroom. After the required number of days of chilling, the meat is cut up and packaged for the freezer.

Animals selected for meat production should be in prime health, not breeding or moulting. The best, or most economical time to kill an animal varies with the species: for example, for beef this is traditionally when the grass gives out in autumn, and ducklings should be killed at about ten weeks or so. (They are much more difficult to pluck when they are older, and eat more than the extra meat is worth.)

Before slaughter, animals should be fasted overnight and allowed plenty of water. Fasting is important because if the blood contains food substances which are only partly assimilated, this may affect the meat. Care should be taken not to alarm or excite the animals: a rise in temperature means that the carcase will not bleed properly.

Resting animals have a reserve of glycogen in the muscles which changes to lactic acid when they are slaughtered, and helps to preserve the meat. So, both for reasons of welfare and meat quality it is important to keep the animals calm and quiet before they are slaughtered.

The meat should be cooled as soon as possible. Poultry and rabbits can be cooled in the fridge, but large animals need a cool room. The temperature right in the middle of the thickest piece of meat should be no more than about 8°C within 48 hours. If you try to do the job yourself, there is a risk of taint with larger animals. There is a lot to lose: a beef animal is worth about $500 alive, and much more when it is cut up for the freezer.

Preserving meat

If you keep pigs, sheep or cattle for meat, you will need to preserve some of it. Freezing is quick and efficient and it inhibits most bacterial growth, although lipolitic bacteria (which attack fat) can live at low temperatures. This means that, in a freezer, lean meat keeps best and beef keeps longer than pork or lamb. To save space, we can bone and roll the meat.

Salting and drying were traditionally used but, nutritionally, they are not as good as freezing. Still, you might like to try your hand at old-fashioned silverside (salt beef) or ham. Before using it, you can soak salt meat overnight, which gets rid of much of the salt. Bacon is another common form of salted meat. You might like to try salting a small piece of pork, to see how it turns out.

Dry salting is simple. It consists of rubbing a ham or shoulder of pork with salt, every day for three weeks. Make a pocket by the bone, and pack it with salt and a hint of saltpetre. The ham is then left to dry, suspended from the kitchen ceiling in a muslin bag, for about one year! Dry salting should be done at a low temperature. Unless you can keep the meat in the fridge for about three weeks, I don't recommend this for Australian conditions.

A brine mixture can be used to 'pickle' the meat. There are many variations on the basic recipe: boil a mixture of water, salt, brown sugar, saltpetre and (if you like) beer. Usually, the meat is left in such a brine solution for about a month, and turned every day.

Using skins

Dressing skins is a skilled job. If you want a soft sheepskin rug for the bedroom, send the skin away to be dressed.

With patience, a rough sort of job can be done at home. The skin must be dried in a cool place with plenty of air around it. Rabbit skins can be dried while pegged out on a board. Rub the flesh side with salt as a precaution.

The dry skins can be stored until you have time to work on them. First soak the skin in clean water until it is soft. Add a little detergent and work the skin about to get rid of blood and fat. Then wring it out and de-flesh it. This is a tricky process: scrape the back of the skin with a paint scraper or knife, removing any remaining fat and meat, without cutting the skin. Soak it again for an hour in water and detergent, and the skin should be clean. To preserve the skin, soak it in a solution of 1 kg of salt and 1 kg of potash alum in 10 L of water. After a few days, drain and slowly dry the skin, folded, again for a few days. Now peg it out flat to dry, away from direct heat. When it is dry the skin will be hard, but it can be softened by working and rubbing with a little oil or fat.

4

The Animals

Choose your animals with reference to their compatibility, the size of your land, the crops you grow, and your needs: make them part of an integrated system of growing, self-sufficient in terms of fertility, and complementary in terms of food.

Bees

If you want an integrated system, sooner or later you will probably consider keeping bees, because they are great integrators. Bees will give you honey while improving the pollination of many crops, and they are being increasingly used in agro-forestry systems with native flowering trees.

Commercial beekeepers normally move their hives on trailers around the countryside to catch the various flowering crops; but with some of the new agro-forestry systems it seems that the bees can settle down in one place and take advantage of a succession of flowering trees and shrubs, which can then be harvested for seed.

Hobby beekeepers, with only one or two hives, usually find that there are enough flowering plants in their area to keep the bees happy all year round. Here are a few pointers before you start:

- You may have an aversion to insects that would make you shudder when handling bees.
- You could be allergic to bee stings, which would make them dangerous for you.
- In the wrong place, bees can be a nuisance to neighbours and to livestock.
- You need more than textbook knowledge. Get to know a beekeeper and handle some bees before you decide to buy the rather expensive equipment. Still interested? Then read on!

Yes, you can expect a few stings, especially at first, but unless you are allergic to them, they won't cause you too much pain. In time you will develop immunity, so that the stings won't swell so much – but there will always be some pain. Protective clothing will keep most bees out, and it would be a good idea to get a strain of bee that is quiet and docile. They vary quite a bit.

Bees are good if you have limited time for your enterprise because you don't have to visit them every day: the work is seasonal. However, you do need to check on them regularly, especially in spring. Two or three hives are quite enough for a backyard beekeeper.

As you will imagine, seasons vary so much that it is impossible to predict what your yield of honey will be. The weather affects the amount of nectar produced, and also the number of days that the bees can work. As a rough average, beekeepers hope for about 45 kg of honey per hive in a season. The market price for honey fluctuates according to the season and is also influenced by exports. There are other products to be enjoyed.

Products

Honey is the best form of sugar to eat, according to the health experts, and it has many uses as a wound dressing and preservative.

Mead is a drink made from honey, thought to be the first alcoholic drink and known to the ancient Greeks. It is fermented in a similar way to other wines, but making mead takes a little longer. The ingredients are honey, water and yeast, using about four times as much water as honey.

Fermentation takes six months, and is best in the warmer months of the year. Then the wine is matured in oak casks until it is two years old. This produces a dry mead, and extra honey can be added to make a sweeter drink.

Beeswax (the cappings cut off the comb) is used to make polish or beeswax candles. Beekeepers collect any bits of wax found when examining the hive, because beeswax is valuable. Brood comb, containing the remains of old cocoons, is also rendered down after soaking in water.

Wax is rendered by heating in water at a temperature of 75°C. The wax can be suspended in a muslin bag under the surface of the water, and when it melts it will come out into the water, leaving the impurities in the bag. Let the wax cool slowly. In a solar wax extractor the sun supplies the heat.

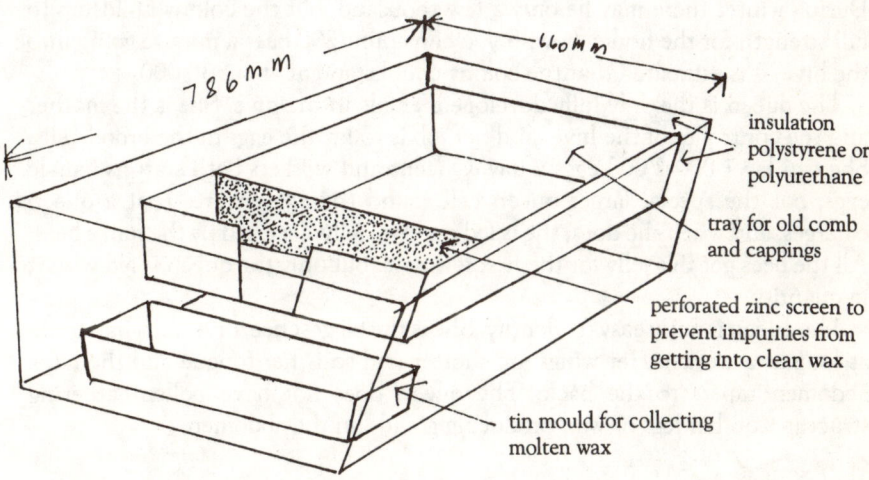

Solar wax extractor   This extractor consists of a box with a double-glazed lid, inside which is a sloping tray. When the extractor faces the sun, the wax melts and runs down through a filter into a trough at the bottom.

About bees

Bees kept for honey production belong to one species, *Apis mellifera*, the honey bee. This is an insect of the warm temperate and subtropical regions, a forest-dweller for thirty million years. Three subspecies are used in Australia. These subspecies can interbreed, so there are variations.

Apis mellifera ligustica, the Italian bee, is the most common in Australia. If you look at the workers, you will find between two and five yellow bands on the abdomen. These bees start working quite early in spring, and they have fairly long tongues with which to reach into flowers for nectar. Their temperament varies, but some strains are bred for docility and predictable behaviour.

Apis mellifera carnica, Carniolan bee, has brown bands of short hair on the abdomen and a greyish appearance. They also start early in spring, but some of them swarm often.

Apis mellifera caucasic, the Caucasian bee, has brown bands, but looks darker than the Carniolan. It has the longest tongue of the three, and can reach into flowers the others can't reach. Caucasians are usually very docile and build up colonies quickly. They are so keen to fill up gaps with propolis that they can make management rather difficult.

Bee biology

A colony consists of thousands of insects, all functioning as a single individual. During winter there may be only a few thousand, but the colony builds up to full strength for the honey flow. If you can count 200 bees a minute going into the hive, it is estimated that the colony could stand at about 30 000.

The queen is the only fully developed female in the hive. She is the mother and the controller of the hive, and her job is to lay the eggs in the brood cells. She can lay 1 000-2 000 eggs a day. Queens and workers both start as female eggs, but the special larger queen cell round the egg ensures that a queen emerges, and when she does, she is fed on royal jelly, secreted by the nurse bees. All the bees get this jelly for the first few days, but only the queen is fed with it in quantity.

The queen is fairly easy to identify. She is the biggest bee, one and a half times as long as a worker. Her wings are shorter and so is her tongue and the large abdomen tapers to the back. The queen does not have pollen-gathering structures on her legs or wax-producing glands on the abdomen.

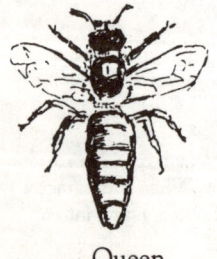
Queen

Worker

Drone

Drones hatch from eggs that are unfertilised, in larger cells than the eggs for the worker bees, and beekeepers remove drone cells when they see them. Drones are male bees, and they do no work. Their only function is to mate with a queen. They are turned out to starve by the workers in the autumn, and do not survive. The drone is a heavier insect than the worker, with a square-ended abdomen, long wings and very large eyes. Drones don't sting, because they have no egg-laying instrument, which is used by insects to sting.

Worker bees are the most numerous, and they are the ones we see outside the hive. Workers are all females, but without fully developed reproductive organs, so they do not lay eggs as a rule. They are the smallest of the three bee 'castes'.

The worker has wings that reach beyond her abdomen, a long tongue and a barbed sting for the defence of the hive. On the hind legs are combs for removing pollen collected on the body hairs. The combs sweep the pollen into a sac on the hind legs. Glands on the abdomen produce wax for the honeycomb in which the honey is stored.

Life cycle

Life starts as an egg laid by the queen, one egg to a wax comb cell. The whitish larva hatches out and is fed by the worker bees. The larva grows fast, shedding its skin four times. When it is fully grown, the workers seal the cell with wax, and the larva spins itself a cocoon within which it pupates. The pupa stage is immobile, and the adult body develops inside the case, emerging when ready to join the colony. The brood is the hive population of immature insects in their cells.

The queen bee can determine whether she lays fertilised or unfertilised eggs. Unfertilised eggs develop into male drones, fertilised eggs into female workers (or into a queen if fed mainly on royal jelly).

The cells vary according to their purpose. Worker cells are smaller than drone cells, which are usually in the lower parts of the comb. Worker cells are also used for storing honey and pollen. Queen cells are built out from the comb, sometimes only one or two. These cells are shaped like a wax cup and hang down from the comb. They are started if the queen is killed or if the bees want to swarm.

A young queen will emerge on the sixteenth day and stay in the hive for a few days, feeding. If she meets another queen, she attacks her. After a few trial flights from the hive, she will go on a mating flight, will mate with a drone and then go back to the hive to start laying eggs. She will not leave the hive again unless she leaves with a swarm to start a new colony.

If the queen's wings are not properly developed and she doesn't fly, she will still lay eggs, but they will be unfertilised and will all produce drones. Beekeepers can usually spot this by looking at the very domed cell caps and the irregular laying pattern.

The workers' first job is to feed the younger ones and the queen, and to help with the chores, such as air-conditioning the hive by fanning their wings (bees can control the temperature and humidity in the hive). They store the nectar brought in by field workers. The nectar is manipulated to get rid of some of its moisture before storage, and after about three days the cell is sealed with wax.

In the cell a change takes place, and the sugar produced by plants is converted into up to thirteen different sugars present in honey, mainly dextrose and levulose. After about three weeks of hive duties, the worker goes out collecting. In the busiest part of the year, workers live for only a few weeks before being replaced by younger bees.

Replacing the queen

Swarming can be a problem sometimes if the hive is overcrowded, or underworked. The bees will then decide to look for a new home. They cluster round the queen and swarm onto some object, such as a tree trunk or a building. Your colony may need more space, or better ventilation to prevent swarming. New queens are less likely to swarm, so some people replace the queen with a new one each year.

Diseases

Disease can appear in either the brood, the larva and pupa stage, or in the adult insects. The most serious are brood diseases.

EFB (European Foul Brood)
This disease causes an irregular brood pattern with dead brood curled in open cells. Larvae are discoloured instead of white and shining. You can usually remove the dead larvae with a match. The cappings may seem normal, but they will be irregular, and they can be dark and sunken as in AFB. The comb may have a bad acid smell.

AFB (American Foul Brood)
If you suspect this disease, you are required to report it to an apiary inspector of the Department of Agriculture in your state. It may be difficult for a beginner to be sure what the problem is when the brood is abnormal, but seek advice. In AFB the brood pattern is irregular, and many dead larvae are found on their backs at the bottom of the cell under dark and perforated cappings. There may be a foul smell and the dead larvae may have formed a black scale on the cell walls.

Bee crops

Your immediate area should be carefully watched, for the species of flower and the flowering times, although these vary from season to season, which is one of the reasons why honey production is so unpredictable. You can then plant species to fill in the gaps, when fewer flowers are available. You will need to learn which flowers have good yields of nectar, and which have pollen, needed by the bees to feed the larvae.

Eucalypts are the most important species of plant for bees, but there are hundreds of species, and very few people will know them all. The nursery I work with sells the following species, which are visited by bees:

E. *maculata* – spotted gum
E. *viminalis* – ribbon gum
E. *citriodora* – lemon scented gum
E. *radiata* – narrow-leaved peppermint
E. *leucoxylon* – yellow gum
E. *ficifolia* – red flowering gum

There are many others: yellow box, red box and grey box, and the ironbarks and stringybarks are also useful. Other species include *Angophora costata* and *A. floribunda*, the smooth-barked and the rough-barked apple. The callistemons are not very important, but sometimes they will keep the bees going until something better flowers. Some of the paperbarks (*Melaleuca*) and the tea-trees are useful to bees.

However, the common tea-tree (*Leptospermum flavescens*) tends to make the bees produce a jelly-like honey, which is good bee food, but not extractable. The wattles are more important for pollen than for nectar.

About eighty per cent of fruit-tree pollination is by bees, so close to the orchard is a good place for beehives. The reason that apple blossom honey is not famous is that fruit trees flower in spring when the colonies are building up, and so the honey produced at this time is used for rearing more bees.

Bees love the aromatic herbs such as melissa, rosemary and thyme; lavender is good for nectar. Fennel attracts them, but the honey has a strong flavour. Bees also like sweet-smelling old-fashioned flowers like wallflowers, and weeds like dandelions and daisies. They love clover, and can often be heard working in it, although red clover nectar tends to be difficult for the bees to reach. White clover does not pollinate without bees, and it yields a good supply of nectar, which makes very good light honey.

Canola is used in margarine production, and may be grown in your area. Bees make a lot of honey from canola, although it tends to be extremely stiff and solid, and colour the honey an acid yellow. I remember beekeepers in England complaining about it and telling me that the honey from all brassicas tends to be stiff and hard to handle. For winter, camellias are a good bee standby.

Starting with bees

When there are too many bees in a hive, the bees swarm, and this is the classic way to acquire a colony. However, it may be years before a swarm of bees comes your way, so many people start off by buying second-hand hives with bees included. The hives should be inspected by an expert for the presence of brood disease, so take an experienced beekeeper with you when you buy bees. You should also ask for a young queen. (Queens are often sold separately.)

Sometimes you can buy a nucleus hive from a beekeeper. These consist of a queen, a few frames with comb containing honey, pollen and brood, and enough worker bees to take care of the queen and brood. This nucleus can be moved carefully into your new hive and will expand to become a normal size colony in time. Keep them in one brood chamber until they need to expand.

While the new colony is building its strength, you will need to feed it to help it along. Sugar syrup can be made by mixing two parts of sugar to one part of water and the mixture is given to the bees in a feeder. Dry sugar is sometimes fed, but more as a winter supplement than to stimulate hive activity.

The aim of the beekeeper is to help the bees to build up strong colonies, and to control swarming in spring: you don't want to lose your bees! Often people feed the bees in spring with sugar syrup to encourage them to get going before the honey flow, or between flows, when there is little nectar available. Honey is collected from the hive in summer, but in the colder parts of Australia and New Zealand the autumn stocks are usually left in the hive to provide winter rations for the bees. The average hive will need about 16 kg honey for the winter.

You will learn from other beekeepers when to feed the bees, when to add supers, when the bees are likely to swarm, and what to do about it.

Basic equipment

Beekeeping equipment is available from 'Going Solar', 320 Victoria Street, North Melbourne, Vic. 3051. Send 4 stamps to obtain their catalogue.

Protective clothing A firm hat, folding wire bee veil, combination overalls, beekeeping gloves. Cost, $90–$100.

New hive components You can make them yourself from factory-made components. Cost, say, $100 plus.

A nucleus hive You may be able to take a swarm in spring and get one free. Otherwise it may cost about $45 for the bees.

Smoker and hive tool The smoker is a hand-held fire box with a set of bellows attached in which you burn leaves or bark to make smoke. Cost, about $45.

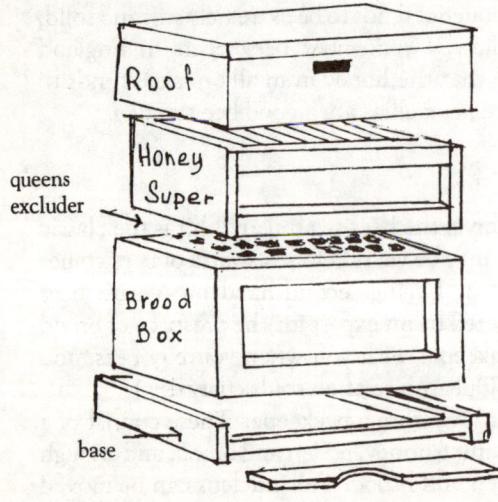

Beehive There are various designs of bee hive but they are similar, based on the Langstroth concept of 'bee space' which is the exact space in which bees build their honeycombs. Moveable frames are necessary under the Bees Act and they allow you to carry out the necessary tasks with the minimum of disturbance to the bees.

The hive consists of a waterproof roof with a ventilator and bee escape in it, under which are the honey supers. Below this is the queen excluder and then the brood chamber where the queen lays eggs for the next generation. At floor level is the hive entrance with an alighting board. The whole thing usually stands on a base to keep it off the ground.

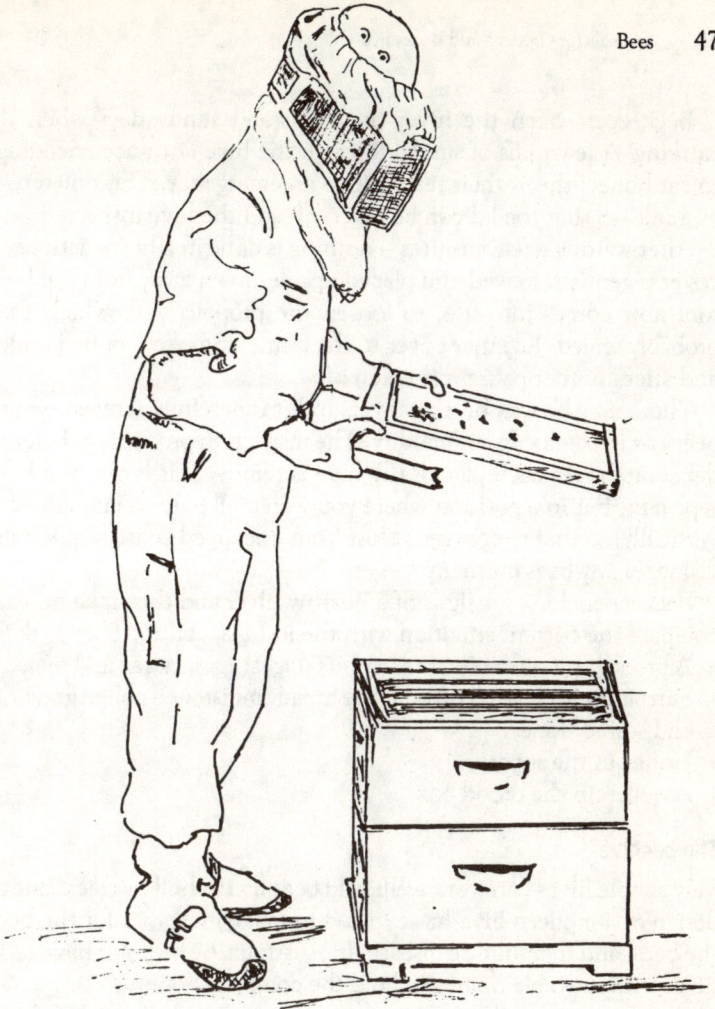

Everyone keeping bees has to register as a beekeeper under legislation aimed at controlling bee diseases. A check with your State Department of Agriculture will give you up-to-date information on this, as well as access to some advice about beekeeping.

Working with bees

Rough treatment is what bees hate most, so be quiet and gentle with your bees, as with all stock. They seem to hate dark clothing and black watchbands: commercial beekeepers wear white, zip-fronted overalls for this reason. Bees hate fluffy material, because they get stuck in it, and then they sting.

Bees also dislike strong smells. On a course I attended, we were encouraged to wash with unscented soap before approaching the bees, as they don't like the smell of perspiration, and to avoid perfume, aftershave and hair spray.

I can remember the excitement of seeing a hive opened for the first time. We stood to the side, so as not to impede the workers as they went in and out of the hive.

Beekeepers open the hives on warm and sunny days when the bees are working. A few puffs of smoke through the hive entrance encourages the bees to eat honey: this is their response to an emergency. This quietens them down as a rule, so that the lid can be taken off and the hive inspected.

After waiting a few minutes – nothing is done in a hurry with bees – the hive cover is gently removed and placed upside-down away from the hive. The hive tool now comes into use, to loosen the propolis with which the bees have probably sealed the inner cover to the frames. Smoke is puffed under the cover, and after another pause it is taken off.

There may be wax on the cover, and it is carefully scraped off and saved, as beeswax is a valuable commodity. The inspection proceeds as before, slowly and deliberately. As each part of the hive is removed, it is put in a handy spot for replacing, but in a position where you won't fall over it. Frames are always held vertically, so that nectar is not lost from uncapped cells. A quick shake usually dislodges any bees on them.

Beekeepers look for signs of a healthy hive, and they take notes so they can compare the current situation with the last inspection. They look for:
- A brood box containing the various stages: eggs, larvae and sealed brood cells.
- Surrounding cells full of food: bee bread (moistened pollen and honey), pollen and stored honey.
- Honey in the supers.
- A queen in the brood box.

The beehive

Very simple hives can work well: wild bees live in hollow trees. But the complex design of a modern hive has evolved to make life easier for the beekeeper and the bees, and to minimise disease. In Australia, by law, you have to keep bees in hives with movable frames to hold the combs of honey.

Our ancestors made beehives from straw in the traditional beehive shape, but modern hives are square. They consist of a floor, a deep box with two or three shallow 'super' frames, an inner cover and a roof. Sheets of foundation beeswax, made by a machine to imitate exactly the pattern used by the bees, are provided to start the bees off with honey storage. This keeps the honey separate from the brood of immature bees, because in a modern hive the brood chamber is below the honey storage. A queen excluder keeps the queen out of the honey area, so no eggs are laid there.

It may be better to buy the parts for a beehive already made and assemble it yourself, rather than to make one from scratch, because the dimensions are critical. The bee space has to be exactly right between the frames for the bees to be happy. But if you are handy with tools, get a proper working drawing from your Department of Agriculture and work exactly to the measurements.

Siting the beehive

To keep your bees out of the neighbours' hair, you need to arrange for them to take a high flight-path in and out of the hive. If you have near neighbours, keep the hives well away from the boundary fence and up against a hedge or other barrier to make them fly high. People with a few acres usually position the hives in a quiet, sheltered place, well away from the backyard. Suburban beekeepers often keep them on the roof.

Until you start to look for them, you won't notice many beehives around. This is because beekeepers find that it pays to be as unobtrusive as possible with bees. If the hives are not noticeable, there will be less danger from vandals, and possibly fewer complaints from neighbours who've been stung by a bee and are convinced it is one of yours. A beekeeper I know used to keep hives at Fountains Abbey, which is the most visited National Trust property in England, but he was careful to keep them well off the beaten track, and few of the visitors realised they were there.

Bees need to live near their work, so the nearer you can place them to sources of nectar, the better. Bees can travel over three kilometres in search of nectar, but on a long return trip with a full load, they can get short of energy and consume some of the nectar. This is why commercial beekeepers move the hives to the crop.

Of course, bees need a source of water close by, with something in it to stand on, such as a stone. Some people provide a shallow container partly filled with sand, and let a tap drip onto it. Bees like warmth and dislike wind, so a protected site is best, perhaps with shade in the hotter months. They seem to like the morning sun on the hive entrance.

Bees hate the damp, so don't put them in a gloomy gully, or place the hive straight on the ground. Some beekeepers make a concrete stand for the hive, sloping slightly so that the hive drains to the front.

Allow plenty of room for access all round the hive for working with the bees. Hives are usually painted white; some people think it makes it easier for the bees to find their way home.

Honey harvest

Nectar is collected by worker bees from flowers and also from nectaries on the stems or leaves of some plants. Nectar is made up of sugars manufactured by the plant during photosynthesis. Conversion from nectar to honey occurs after storage in the hive. The enzyme invertase, which comes from the mouthparts of the bee, is mainly responsible for the conversion. When the conversion has taken place and the honey is ripe, the cell is capped with wax by the bees. Honey keeps well in the capped cells, so it is easier to wait until there are several capped frames ready before taking the honey. When a frame is full, you can add another super, under the full one. Your honey crop is the excess honey which the bees store for later use.

If you are too impatient and take the honey before three-quarters of the cells in the frame are capped, the honey from the uncapped cells will add moisture to the mixture and the resulting honey will ferment.

Taking the honey

Bees are either shaken out of the frames or sent down to a lower level by the use of a chemical repellent like phenol or benzaldehyde. These are dangerous substances, and you may prefer not to use them. You can also use an escape board, which has to be left in position for a day or so. The bees leave the super through a 'bee escape' but can't get back to it. When you return, it is clear of bees.

Full frames are taken from the hives, and the extraction process can be done in the kitchen or a bee-proof shed. Bees will be able to smell the honey, and may come in if they can. The room should be warm (27-30°C), as honey is hard to handle at low temperatures.

Wherever you work, the room and the utensils must be clean, and you should wear a clean overall, as this is food preparation. The work tends to be sticky, so it helps to have hot water ready to wash your hands.

The cappings are sliced off with a hot knife. Keep two large knives in hot water on the stove, and dry each knife quickly before use. (You can buy electrically heated knives for this job.)

For uncapping, you can rest the frame on a wooden slat across a wide basin to catch the wax cappings and any escaping honey. Cut down the vertical face of the frame, deeply enough to expose the honey. The cappings themselves will contain honey, and this will drip out if the cappings are hung over a basin in a muslin bag. A honey extractor is a drum with racks inside it in which to place the frames. You can buy one, but you may be able to borrow an extractor from a beekeepers' association. Some people make their own.

The handle on top of the drum is turned, and the honey is extracted by centrifugal force. It runs down the inside walls of the drum and can be filtered as it runs out of the bottom into a storage vessel. The speed has to be just right: too fast and you damage the combs, too slow and honey is left behind.

People without an extractor scrape the honey and wax down with a spoon into a muslin bag, from whence the honey seeps out, leaving the wax behind.

The honey will be cloudy at first, with air bubbles and bits of wax in it, but these will rise to the surface and the wax can be skimmed off. If necessary, filter again. Store your honey as soon as possible in airtight jars, because honey attracts water, and if it is diluted it may ferment.

The combs will still be wet after extraction, and they are often stored wet, because the bees are encouraged to store honey in them when they are returned to the hive.

Heat up the cappings in a warm oven, so that the wax melts and forms a sheet, with the honey underneath it. This honey can be used for cooking.

Cows

A cow is quite a large animal. It would not be at home in the average suburban block, but for bush blocks and where milk is not readily available, the cow can be the centre of the enterprise and a valued colleague and friend. Cows vary a lot in size and temperament.

The Friesian

The Friesian is the number one milk-producer all over the world: she is a big, placid milk-machine. Friesians expect large quantities of food and an easy life, with no foraging and no bad weather. They are not memorable for personality, and although some old Friesians settle down to become house cows, there are other breeds that fill the role better.

The Jersey

The beautiful Jersey was once the main source of milk and butter in Australia and New Zealand. They were able to survive a long journey by ship from the other side of the world, so they were the cows brought by the first settlers. The Jersey still makes a good house cow, although the milk is rather rich for modern diets. Jerseys are small and docile, fine-boned, golden animals with large eyes, the most ornamental of all the cow breeds.

The Dexter

In the eighteenth century, Mr Dexter, agent on a large estate in Ireland, bred from some cattle with dwarf genes so that local people with an acre or two of land would be able to keep a cow.

The Dexter is a small, black, hardy animal, which produces milk and a good beefy calf (Jersey calves are not really beef material). But Dexters have come a long way from their peasant origins on the poor soils of Ireland. They are so fashionable at the moment that they could be too expensive, unless you regard your cow as a financial investment. The cows can use poor land and need only 60 per cent of the amount of feed other cows eat.

Dexters tend to be independent little animals, and can be aggressive at times, but if they are handled well, this should not be a problem. The main problem at the moment, as with any fashionable breed, is the price!

The Murray Grey

It was a white Shorthorn house cow that helped to start the Murrey Grey breed at the turn of the century. A farming family on the Murray kept a valuable Aberdeen Angus stud, and every year they bred a calf, by an Angus bull, from their white house cow. Every year the calf turned out the same, a beautiful, docile silvery-grey heifer. They were so attractive that the Murray Grey breed was established from them. From Australia the Greys have spread round the world.

Crosses between breeds do not usually 'breed true', but revert back to one side or the other. However, Murray Greys are now a fixed breed with a certain amount of colour variation, from mulberry to almost white. The Murray Grey is probably the most docile cow in the world! Other milk breeds include the Guernsey, a Channel Island breed similar to the Jersey; the stylish Ayrshire; and the Illawarra Shorthorn, an Australian version of the old Shorthorn breed.

My favourite animal for a house cow is a Murray Grey-Friesian cross, of which we have several on our Gippsland farm. We bought these animals as calves from a dairy farmer who had used a Murray Grey bull on his Friesian herd. We reared them ourselves, they became very friendly, and when the time came, they calved without problems. The combination of breeds means that these cows have plenty of milk, and also produce a smallish but solid beef calf when mated back to an Angus bull.

Getting started

All the breeds of cattle have their breed societies, so information is easily available. Some of the societies have their own magazines, and breeders advertise in stock-farming magazines. The breeds of dairy cows, kept mainly for milk, are used to a close relationship with human beings, and they are usually easy to handle.

It is probably easier for beginners to buy a mature cow, used to hand-milking. A clearing sale is probably the best place, where a whole herd is being sold. An experienced farmer is the best person to go with you to the sale. Another way would be to buy a heifer calf of a dairy breed and rear her to become a house cow at between two and three years, when she has her first calf. This will be cheaper, and will ensure that you know each other when the business of calving and milking begins.

You should get about 3 000 litres of milk from a cow in one lactation, which is quite a lot more than the cost of feeding her. A healthy calf each year should be saleable, or could be reared for beef, or for future milking if it is a heifer.

The value of the calf will vary widely, according to the breed, the area and the prevailing prices. But if you want to sell a good calf, use a bull of a beef breed. The Aberdeen Angus is usually used on small cows and heifers, while the Hereford is popular as a cross with larger cows. Herefords, with their large, white faces, are very good beef animals, but I wouldn't keep a Hereford for milk-production because it is not their strong point.

Feeding

Cattle, like sheep and goats, are ruminants, and their main food is grass, so in areas of good rainfall you should be able to feed the cow for most of the year on between one and two hectares of paddock.

One of our Murray Grey cows, with her Angus cross calf.

Rumination is an involved process, and if it stops, the animal is in trouble. 'Is she still ruminating?' is the first thing we ask when we suspect a health problem. It is usually possible to tell if you keep an eye on her over a period of a few hours.

The ruminant has four stomachs. Grass is swallowed straight down, without chewing, into the first stomach, the rumen. This is a big pouch with a capacity of about 200 litres, a mixture of grass, moisture, and methane gas – of which more later. When the cow has finished grazing, and is at rest, she chews the cud for about eight hours in each twenty-four hours. She regurgitates a 'bolus' of grass back into the mouth and chews it slowly and thoughtfully before it goes back down again for further digestion. The food is thus finely ground, and is ready for the bacteria to work on it. There are many intestinal bacteria, and the type depends on what food the animal has been eating, which is why a ruminant's diet should be changed gradually.

As the food is broken down by bacteria, fermentation takes place, and methane is produced. The gas is usually belched harmlessly away, but when the animal 'bloats' the gas is trapped and can kill the cow by pressure on heart and lungs. After the muscular churning in the rumen, partly digested food goes on. The second stomach is the reticulum, where foreign objects are sifted out and retained. Nails and wire have been found in a cow's reticulum, but there is obvious danger in letting an unselective grazer like a cow into a paddock with such hazards about. It is very important to make sure that, when you have finished fencing, all objects like wire and nails are removed. If you don't, grass will soon grow over them, and they will cause trouble later.

The omasum and abomasum are the third and fourth stomachs, the abomasum being the true stomach, where digestion proceeds as it does with non-ruminants. The food substances are broken down by enzymes (starches and sugars into glucose) and enter the bloodstream through the stomach lining.

Bacteria in the rumen can break down fibre and cellulose. If there is not enough bulk in the ration, rumination is affected and digestive problems occur. I have seen cows suffer from ketosis on a winter diet of too much barley meal. Always make sure that ruminants have plenty of 'long fodder': hay, silage or grass. In times of grass shortage, when you may have to feed grain, straw makes a useful roughage. Although straw has little feed value, it keeps the rumen going.

A cow needs a lot of bulk, the equivalent of about 12 kg of hay a day for maintenance and some milk-production. It is possible to keep a cow on hay plus surplus vegetables, weeds and so on, if you are short of grass. To calculate how much hay you might need in addition to grass, work out how many days there are when grass does not grow during an average winter period, or during the summer dry period in the north. Allow 12 kg per day for that period. In the Gippsland hills, for example, we have a 100-day winter as a rule. Of course, seasons vary, but you have to take an average in order to be prepared.

Watch the body condition of the cow, because it will tell you whether she is getting enough to eat or not. The degree of fatness will depend to some extent on the breed, and most dairy cows tend to be bony. However, a very thin cow will not milk well, and she will always be hungry. Make sure your cows are well fed.

Various supplementary feeds can be used for cattle, to keep up the body weight of a good milker, or to put weight on a beef animal. These can be either grain or by-products of the food industry. In England we used to feed sugar-beet pulp, and the cows loved it; this was a by-product of beet-sugar production. Similarly, in sugar-cane-growing areas of Australia you can get molasses quite cheaply, and this makes a high-energy feed supplement. Then there are brewers' and distillers' grains, waste-products from beer and whisky manufacture.

With our cattle, pollard is popular. This is a wheat product, with a little more feeding value than bran. If your cow is too thin, a little dairy meal will probably build her up again, a mixture of grain with protein added. This is the food used by dairy farmers to boost the yields of cows, and if you expect Buttercup to have a high yield, she may need this sort of help.

How much milk to expect from your cow is almost a philosophical question. Commercial dairy cows give a great deal of milk, about 5 000 litres per lactation. This puts a lot of stress on their systems, and it is quite usual for a dairy cow to be worn out after about four calvings. Beef cows will live longer productive lives (about ten calves is the average), and the house cow, if not pushed to produce the maximum yield, will with a bit of luck be with you for many years.

Stress for commercial cows includes not only giving 25 litres of milk a day, but also living in a large group, sometimes too large for establishing a hierarchical order. Cows don't mind walking long distances for food and water, but if you are carrying 20 litres slung in the udder plus a 20 kg calf in the uterus, it is a bit harder. And paddocks without shade from the sun or shelter from wind and rain are stressful too.

Cow needs

The cow's simple needs are quiet handling and consideration. A cow would ask, if she could, for:
- Plenty of roughage for her rumination process, either grass or hay.
- Good clean water to drink.
- Shade from the sun and shelter from wind and rough weather; perhaps under thick bush, not necessarily a shed.
- A quiet place in which to be milked (and from your point of view, it should also be clean).
- Not to be hurried too much from one place to another.
- A mineral block, if the land is mineral deficient.
- Company of one sort or another. Sheep or goats, or even pigs, can be company for a sole cow.

The cow would much prefer to rear her calf rather than have it taken away from her. There are several compromises that can be reached with a house cow, so that you get some of the milk and the calf gets enough too. You can segregate calf and cow at night in a shed, where they can see each other but the calf can't drink. In the morning, milk out your share before the calf gets breakfast. Or you can milk out what is left after the calf has fed.

Dairy cows have the calf taken away soon after birth, and some are quite placid about this, since the dairy breeds have been selected for milk-production rather than maternal instinct. Beef cows seem to be much more maternal and possessive about their calves.

Cow health

The most common health problem with dairy cows is mastitis or inflammation of the udder, which is extremely painful for the cow, and which produces infected milk. House cows are not so likely to get mastitis as commercial dairy cows, because they are not stressed in large herds or pushed for maximum yield. But they can become infected, and treatment should be immediate.

The immediate causes of mastitis are specific bacteria, but the underlying cause may well be stress. The first sign of mastitis may be a hard, hot udder, or a few clots in the milk. Treatment is usually with antibiotic on prescription from the vet, so you need expert advice. Sometimes a mild case can be cured by frequent milking out by hand of the affected quarter and bathing with warm water. The disease usually affects only one quarter at a time.

The question of antibiotic use is often brought up by people who want to keep to organic principles. My belief is that antibiotics have their place as infrequent treatment of serious conditions. They certainly save a lot of suffering both for people and animals. I think we should use them sparingly, and always try to work out the underlying cause of the disease condition, so that it can be put right. House cows should be tested for TB and brucellosis, or they could be a health hazard to your family.

Adult cows usually don't receive regular medication. Hooray! So organic milk is a distinct possibility. The problems they do suffer tend to be related to minerals in the diet, and a mineral lick might help prevent them. Milk fever, for example, causes cows to collapse just before or just after calving, and is caused by calcium deficiency in the blood. Magnesium deficiency causes 'grass staggers'.

Bloat

Another problem to watch for with any ruminant animals is the dreaded bloat, which some people think is also mineral-related. Bloat mainly occurs in a wet spring and on fast-growing clover pastures. The contents of the rumen gas up, and unless the pressure is relieved, the animal dies.

Reducers of surface tension, such as oils, are used for bloat prevention and treatment, but in an emergency the side of the animal has to be pierced in the right place to allow the gas to escape.

Milk fever

I have seen a lot of this in commercial dairy cows, and it could alarm you very much if it happens to your house cow, but this is a metabolic disorder and is usually quite easy to reverse. Milk fever isn't a fever, it is more like a coma caused by a sudden drop in blood calcium. The cow sways, paddles and then goes down with her head characteristically turned to the side. An injection of calcium borogluconate will soon put her on her feet again as a rule.

This disorder usually happens soon after calving, when the milk begins to flow, so one method of prevention is not to milk out the cow completely for the first few days after she calves. You can buy a pack of calcium borogluconate with which to treat a cow with this problem, but it is better to call the vet. Untreated cows may soon die, however, so a pack is a useful standby. Vets normally inject the substance directly into the bloodstream of the animal, but farmers tend to inject it under the skin. This is less risky, but of course its action is slower. If by chance your diagnosis is wrong, calcium under the skin shouldn't hurt the cow.

Foot infections

Foot problems are fairly common in commercial cow herds where cows have to walk long distances to pasture. House cows can sometimes get them too. Infection between the claws of the foot can be caused by stones that break the skin and allow bacteria to get in. Professional help will be needed. The vet may prescribe an antibiotic spray. The old-fashioned remedy was a bran poultice, and people spent a lot of time making 'boots' for cows so as to keep the poultice on.

Ongoing treatment could be a footbath containing either five per cent copper sulphate or ten per cent formalin, similar to the treatment used for sheep. The cow stands in the footbath and the claw opens up to allow the solution to penetrate. Keep an eye on the health and comfort of the cow all the time. No problem should develop to the point where it becomes severe.

Breeding

Most cows are mated once a year, and the gestation period is nine months. After calving, they milk for about ten months and then have about eight weeks' rest before calving again. Some dry themselves off, but some have to be dried off for their own good. To do this, take the cow off any supplementary feed and milk her once a day only, instead of the usual twice. She should be off grass and on hay and dry feed for a few days until the udder goes slack and the danger of inflammation is over.

The cow should come into oestrus or heat every twenty-one days, until she is pregnant. If you are used to oestrus in other animals, you will probably recognise the signs: restlessness, bellowing or discharge from the vulva are all possible. Sometimes it is hard to tell when a single cow is in oestrus, but when she is with other cattle, they will often mount her; and when she is on standing heat, she will allow this.

The best time to get your cow in calf again is the second or third time she comes on heat after calving. This will probably be from six to twelve weeks after she calves. The aim is for a calf every year, so by three months after calving she should be in calf again as a nine-month pregnancy will see her calving just within a year. If you buy a cow in milk, she may be in calf again. Be sure to find out whether she is and, if so, to what breed of bull. Even if she's supposed to be pregnant, look out for signs of oestrus in case she is not.

The easiest way to get the cow in calf is to use a neighbour's bull, but you should be sure that the breed of bull is right for your cow. Cows can have calving problems if the calf is too big. In general, a heifer having her first calf should be calving to a bull of a smaller breed, such as an Angus or Jersey.

The European beef breeds, such as Simmental and Belgian Blue, are popular now with some farmers, but steer clear of them for your Buttercup. The calves of these breeds are very large, and often the cows need help at calving – sometimes they need surgery. I don't think they mix well with the smaller breeds.

I like the Aberdeen Angus, a hardy, little, black animal, for beef, and Angus bulls are often used by dairy farmers. You can never be completely sure of an easy calving, but black Angus calves are usually born without too much trouble. They are small at birth, but sturdy and hardy little things.

Having found a bull of the right breed, take the cow along at the right time, and then *check* after twenty-one days, because if she comes into oestrus again, she isn't in calf. Remember to record the date of mating, and work out when she is due to calve (284 days on), so that you can keep an eye on her at the right time.

If the cow is on dry fodder, try to make sure she gets laxative food when she is due to calve. Fresh, green grass is the best food for cows. In a dry season, you could give her a mix of bran and molasses. But don't give her too much food just before calving. A little less is probably better.

It is not always easy to tell when a cow will calve. Sometimes you get warning, but other times they will make you think it is imminent for days on end. As calving time gets near, the udder will fill up and the pelvic bones drop a little. A cow about to calve may be restless, pacing up and down, and may wander away from her companions. Once she starts straining, the birth has begun, and if nothing happens after about an hour or so, call a vet or experienced handler.

Most cows calve without help, but sometimes a calf may be in the wrong position for birth and needs to be turned before it comes out. The right position is when you can see two forefeet together, pointing down, with the nose lying on them as if the calf is going to dive out.

The cow should be encouraged to lick the calf dry, but she needs peace and quiet to adjust to her new situation. Soon the calf will be up and looking for its first drink. Don't hang over them too much, but watch discreetly to make sure the calf feeds within a few hours.

Calf rearing

You may decide to rear a bought-in heifer calf for milk-production when she matures, and you may want to rear the calf your Buttercup produces. This is not too difficult. It is sometimes easier to milk out your house cow completely and then to feed the calf from a bucket or from a teat with a tube into a bucket. Colostrum, the first milk that all mammals produce after the birth, is vital to the calf's health. It contains a lot of energy and protein and also antibodies to protect the young animal from disease. The new calf needs colostrum within a few hours of birth, and it should continue to get its mother's colostrum for at least four or five days. This milk is thick, rich and deep yellow. Most people leave the calf with the cow at least for the first few days, and during this time it drinks frequently. Calves learn to put down their heads instead of tipping them back when they drink. However, the teat into a bucket is popular because if you rig it up high, the calf tips its head back, which is more natural.

I used to teach calves to drink straight from the bucket by getting them to suck my fingers and then gently lowering my hand into the milk, with the calf still sucking. Don't go down too far into the bucket, or the calf's nose will be submerged and a lot of spluttering will result. A calf can be frightened by this, and will be much harder to teach as a result. After a few tries, the calf will learn to drink without your fingers in the milk.

A calf will need about five litres of milk a day, rather less to begin with. Milk is usually fed in two serves, morning and evening. Smaller breeds, such as the Dexter and Jersey, will obviously need less milk than a Friesian or Hereford, but go by the size of the calf rather than the breed.

Warm milk, at blood heat, is best for calves, but if you are using milk that has cooled, you can warm it up by adding hot water. It is sometimes just as well to add a little water to rich milk such as that produced by Jersey cows. If you segregate the calf from the cow, keep it in a clean pen with plenty of water and deep straw bedding. Allow it a little net of hay (like the one shown for rabbits) and calf pellets, to encourage it to eat solid food.

Calves can be weaned from milk when the rumen has developed. The rumen doesn't work in the new-born calf. Three months on milk used to be the traditional time, but dairy farmers now wean their calves earlier. Owners of house cows leave the calf on milk when they have plenty of milk, and think about weaning when they need the milk elsewhere.

Beef calves are usually weaned when they are about nine or ten months old, and at this stage there are not too many regrets on either side. If you decide to go for home-produced beef, a long period on milk would give the calf a good start. If you are sure the calf is eating solid food and chewing the cud, slowly decrease the amount of milk it gets and increase the solid food: it should be eating grass by this time in a nice little calf paddock. Drop milk feeding to once a day, and then phase it out entirely.

My brother, feeding a Friesian calf in the days when we had a dairy farm. The calves were left with the mother for a few days before she went back into the milking herd without too much anxiety - Friesians are not very maternal.

We reared the calves on whole milk when we had it. This is the best way. They got two or three feeds a day of warm milk in a bucket. Clean water and hay were always available. Soon they started chewing hay and calf pellets, and their rumens started working.

The calves were grouped in twos and threes, in big clean pens. In good weather they went outside during the day. The calf pens were dismantled and scrubbed after each batch of calves and we had very little trouble with calf health.

Cow's milk is the best food for calves. Powdered milk substitute never seems to work as well as the real thing, and the last time we used it we decided 'never again'.

Scour in calves is the worst problem as a rule. The immediate causal organisms are the 'bad' bacteria, but they are triggered off by stress. Bad weather, irregular feeds, any kind of stress can lead to calf scours or diarrhoea.

When a calf scours, the dung will be white or yellow and very runny. The main thing is to prevent dehydration and to replace the bacteria of the intestine with more favourable ones. Vets can now supply you with probiotics, which are like a form of powdered yoghurt, acid-producing bacteria. Or you can use live yoghurt instead. There are solutions, called electrolytes, which assist with rehydration.

Beef

There is always the question of what to do with the house cow's calf. If it is a heifer, you could rear it or sell it as a calf to somebody who wants a house cow. But the male animal is for beef, either for you or another family. It is hard to eat beef when you have known the animal personally, although some families manage it without a problem. If you want home-killed beef, there are now travelling butchers who will bring a coolroom onto your property, kill the animal and prepare it for the freezer. This is a most humane way of treating the animal, much better than the conventional one of sending it to a slaughterhouse, sometimes through a saleyard first. The kindest way is on-farm killing, so don't think of it as barbaric.

Selling beef can be either as a 'store' animal for someone else to fatten, as a 'vealer' or young beef to the butcher at about one year old, or as a fat steer at about eighteen months to two years old.

Suckler cows

An alternative to the house cow is the suckler, a cow that is not milked but which feeds her own calf; sometimes two calves or more if she has the right temperament. Suckler cows are often a beef breed or a beef–dairy cross. They look after their calves very well, and the calves grow fast and have a wonderful life. If you like cows, but don't want to be tied to the chore of milking, a suckler cow could be the answer.

The calves can stay with their mother until they are about ten months old, which would be the natural weaning time. After this, they can be parted without too much stress on either side, to allow the cow to have a rest before she calves again.

Dogs

Dogs are our allies and companions. The right dog can make a real contribution to your own ecosystem, as they have done for thousands of years in human families.

However, dog ownership is a great responsibility. In a way it is a greater burden than ownership of farm animals, although it is usually undertaken much more lightly. Dogs need people (though people are often too busy to be with their dogs), whereas sheep mainly need other sheep, with a little bit of help from the shepherd.

There are simple rules for successful dog ownership, starting with the right breed for your circumstances. Get the right breed and half your worries are over. After this the rules are:
- Keep control. Fence the garden or another area for your dog to be outdoors and secure.
- Attend to grooming, hygiene, worming, vaccination and the necessities for dog health.
- Follow the local laws regarding collar, name-tag and licence.
- Train the dog in good manners, for your sake and everybody else's.
- Get it used to weekends away from you, so that it is not too unhappy without you.

Choosing the right breed

The right breed will depend on where you live, who you are, and what you would like your dog to do. Large dogs obviously need plenty of space, lots of exercise and large amounts of food. Self-willed dogs need firm owners, small timid dogs need gentle human companions, and working dogs are better with work to do. Many people buy a dog for its looks, current high fashion rating, or because they know the breeder. These dogs often come to grief.

Shepherding and cattle dogs are useful if you have flocks and herds and the patience to train the dogs to work. They make good guard and guide dogs too, but they need robust owners with strong personalities who can dominate them as a pack leader would in the wild. Examples are Collie, German Shepherd and Maremma (about which more later).

The Australian Kelpie has been described as a tough little dog with a lot of guts and enormous stamina. They were developed to work with sheep and cattle in Australian coditions: heat, dust, long distances and large mobs of animals. They are pleasant, faithful companions and good watch dogs too. Their small size makes them suitable for small holdings as well as big properties. But Kelpies love to work, and are a bit lost without it.

Terriers can be rather aggressive. They were used for vermin control. They had to catch rats and smell out foxes, digging them out of the earth or showing people where to dig. On farms they are still useful for keeping down rats and mice, but they are mainly companion dogs these days, especially the Australian Terrier. The Jack Russell, popular all over the world, is a bold little dog originally bred by the Rev. J. Russell in England.

Hounds and greyhounds were bred to pursue game and were mainly owned by the ruling classes in the strict medieval hierarchy. In some countries, peasants were forbidden to own hunting dogs in case they were tempted to poach game. Hounds were often bred to look alike so that the pack was even. The larger breeds were used by people on horseback.

The Bloodhound is one of the larger breeds, descended from an ancient line of dogs taken over to England by the Norman invaders. Beagles are a smaller version of the hunting dog.

The methods of hunting differ with the breeds: Greyhounds use their eyes and see game, whereas Bloodhounds put their noses to the trail. Hounds are used more to find the game than to seize it.

Strangely enough, hounds are gentle and affectionate (their owners call them hounds, never dogs), rather aloof with strangers, and they like to think for themselves. Obedience is not a strong point with any of the hound breeds, as they have been bred to work independently in finding game. And this can make them difficult to manage in a small area.

Gundogs evolved from hounds and many of them stand still when scenting game, such as the pointers. Gundogs find game and, when it is shot, they retrieve it. The great thing about this group of dogs in general is their obedience. They were developed by crossing hounds with shepherding dogs, and they retain both instincts.

Labradors and Retrievers need very little training, because their work is bred in them, but they do need early reinforcement of their natural obedience. They also need a lot of supervision when young to make sure they do not chase game or poultry.

The German version of the pointer is described as 'energetic, friendly, obedient, faithful', and the one I know best is all that, but rather neurotic as well. Labradors are probably the most placid of the gundog breeds. Large companion dogs can be good house dogs if you have plenty of room. People living alone are probably the best owners for dogs like the Chow, that attach themselves to just one person. Sedentary dogs, such as French Bulldogs, are pleasant companions for elderly people, since they like a gentle stroll. The highly energetic Dalmatians are good for young people.

Small dogs, like the Maltese terrier and Miniature Poodle, are very expensive to buy. They are in great demand, because they fit in well with a family, even in the city. They are easy to please, take up less food and space than the bigger dogs, and are usually friendly companions for the whole family.

Guard dogs seem to come in various forms, because guarding is natural to dogs. They are possessive, and your possessions are part of their territory. In a world that seems to become less safe every year, guard dogs are becoming more popular. Think of guard dogs, and certain breeds come to mind, some of them ferocious-looking creatures (like Rottweilers and the Rhodesian Ridgeback, which has permanent hackles), which deter the criminal from the start.

The Mastiff was a guarder of castles and dungeons, like the Bloodhound, and the Dalmation protected coaches from highwaymen. The Dobermann was bred in Germany by Herr Dobermann specifically for protection, but the British police now prefer the German Shepherd for their work. Dobermanns apparently get bored with the routine, and can't stand the cold so well. Although some guard dogs are dangerous, this is due more to their treatment than to the animal's intrinsic nature. Guard dogs have in the past been treated badly and chained up for most of the time, then left to roam factories alone at night. Good guard dogs have a handler and are trained to protect the handler.

A survey of dog owners in Australia has found that the most important functions of the dog are companionship and protection. City-dwellers go jogging with a dog, which is good for both of them, and gives protection to the jogger.

Country folk can leave their utes in town, with tools and purchases on board, because the Heeler keeps everyone at bay, rather like the special dogs kept on barges in Holland to guard loads that can't be locked up.

Our family had a most amiable Labrador which was well trained as a gundog, but had no guarding experience. When we started a milk round, and my brother took the milk van to the local town, the Labrador went with him and guarded the milk and the money box, politely but firmly. Anyone who ventured near, uninvited, found themselves face to face with a set of white teeth and a growl.

Family conference on getting a dog: questionnaire
- Why do you want a dog?
- How much time will you have to spare for the dog?
- Does all the family agree that this is the breed for you?
- Who will house-train the pup?
- What about holidays?
- How much space for exercise is there?
- Do you want a pedigree or not?
- Will you breed from the animal in the future?
- What gender? (A dog may need firmer handling; a bitch will have to be shut up when in oestrus.)
- Would it be possible to take a stray dog and give it a good home?

The conference over, the breed decided and the breeder picked out, you have to choose the pup from a litter of squirming bodies. How do you do it? The experts say you should pick a bold pup, but the pup may well pick you, by running forward to greet you. Seven to eight weeks seems very young for a puppy, but this is the best time to take it home since younger pups adapt more quickly to new surroundings.

Having chosen the pup, go home and get ready for the new arrival. A puppy needs:
- A box or basket with newspaper in it.
- A wrapped hot water bottle for the first few days, to replace the warmth of the litter.
- A very small run.
- Plenty of attention. (Let it outside frequently to encourage house-training from the start.)
- A toy. (Ox hide chewy toys are good for pups.)
- The same diet it had at home.

Make changes gradually, giving five meals a day at first, then letting the frequency of meals decline as follows:
- 3–5 months: 4 meals
- 5–9 months: 3 meals
- 9 months and over: 2 meals

Puppy food at first. Finely cut up food, cereals and milk for the first four months, then stale wholemeal bread. Protein diet, but with plenty of roughage as the dog grows older and can deal with it.

Adult dogs need their two meals a day. Most people give the main meal in the afternoon or early evening, but not just before exercise. It is easy to feed all dried or canned food, but very boring. How would you like it? Try to give the dog variety, with fresh meat and bones. Of course they love people food, so table scraps can be a treat for the dog. Fresh drinking water is essential.

Consistent training is important. All should agree what the dog is not allowed to do. It should have its own quiet place in which to be, with a clear idea of what you want from it. Don't beat it or frighten it, but be firm with the rules. Pick it up and put it outside if you see it urinate indoors, but not if you find it afterwards. It is too late then. Use your tone of voice to let the dog know you are pleased or displeased, and be consistent. (And if you think this sounds like a recipe for bringing up children, it might work well with kids too.)

Most dogs get fleas, so there is a battle to be fought. Grooming is needed for most breeds, and some need nail-clipping at intervals. Exercise is most important. Obese dogs do not live very long. Try to spare an hour a day for exercise with the dog. It will be very good for your health as well. Go somewhere quiet where it can be let off the lead for a frolic.

Farm dogs get more exercise than their city counterparts, but you can't let a country dog wander. It can get into all kinds of trouble, chasing livestock and so on. Many country people find that they have to leave the dog on a long chain for most of the day, letting it off for exercise as much as possible. If the chain is attached to an overhead wire, the dog's range will be large. It must always be possible for the dog to reach water and shelter from extremes of weather, so the kennel must be within range.

Breeding

Dog families should always be planned. It is unethical to allow unwanted animals to be born, but extreme vigilance is needed if you have a bitch in season and don't want her to breed. Spaying (neutering a female) is the wisest course, and is usually done at about six months, unless you hope to breed from your female at a later date. Responsible owners have their dogs neutered. Consult your vet about it.

Just how difficult it can be to control dog breeding was brought home to me recently when I met an accidental litter of puppies: pure bred, fortunately, because they are working dogs and will probably find good homes. The parents had mated through the wire mesh of a pen that was supposed to confine the female in season, but which obviously had not been effective.

The time at which bitches come into season varies. Ours was well over a year old, but most seem to start the cycle at about nine months of age. At this stage they are a bit young to breed and in fact most animals are not bred at the earliest time it is biologically possible to do so. Wait until the second or later season before allowing her to breed.

The bitch will probably be in season for three weeks, attracting male dogs for the whole of this time. This can be trying if you have a lot of doggy neighbours. However, she won't mate except for about a week in the middle of her oestrus period, so you have some warning before the danger period.

A veterinary friend said he had seen some success with a 'chastity belt' for bitches, made of an old rubber tyre. The bitch could still have exercise, instead of being locked up in a shed. The design would depend on the size and breed of dog. When we discussed the problem of a bitch in heat, the vet said that some people ask for a hormone injection for the bitch every six months, which suppresses oestrus. There is also a tablet (megoetrol acetate) to be given early in the heat period. Signs of oestrus are pretty obvious in the bitch. The vulva will swell and you may see a blood-stained discharge, and the animal will be restless and more excitable than usual.

The gestation period is 63 days, and about halfway through this period the bitch will start to show signs of pregnancy: enlarged mammary glands and a swollen abdomen.

It is important, as with all stock, to make sure that a breeding bitch does not get too fat, as this will make whelping (the proper name for dog birth) more difficult. A pregnant bitch will probably not need any extra food except perhaps for the last week or so before the birth, although of course her diet should be nutritious. Afterwards, she will certainly need more food to help her to make milk for the pups.

Bitches can usually give birth without too much trouble, and may prefer to be left alone, but you will need to keep checking that she is making progress. Don't be alarmed if there is a gap of up to thirty minutes between pups; but if she strains for a long time without giving birth, get help.

As you would with piglets, leave the little pups alone if the mother is attentive, but if she is preoccupied, a pup may need rubbing down to dry its coat and get it going.

Maremma guard dogs

This is an old Italian breed from the plains of Maremma and the mountains beyond, where flocks and herds were in danger from wolves and bears. The dogs adopt the flock of sheep or goats as their family, and protect them from predators.

The Maremma is a large, white fluffy dog, with great intelligence, which forms a bond with its adopted family as a pup. This bond is more important than the link to the owner. As soon as it is weaned, the Maremma puppy is housed with one or more animals of the species. This type of bonding is not unusual between species, and we have, in fact, a dog which played with calves when she was a pup and is still friendly with cattle.

According to an article in *Town and Country Farmer*, Maremmas were first imported to a goat farm near Melbourne because kids were being lost to foxes and dogs. Before there had been time for training, the imported dogs befriended the goats and killed a stray dog, which was harassing them. Some farmers have not needed to shed their goats at night since they bought Maremmas.

It is important to feed your Maremma and make it feel that you too are a member of the clan. But owners are warned not to be too friendly with their dogs, unless it is the homestead they are to guard.

These dogs are not usually given the normal obedience training. In fact, they are not always instantly obedient, because they like to think for themselves. An American study of guard dogs, which went for 10 years and covered 100 dogs of five breeds, showed the Maremma to be the best guard dog. Sometimes the presence of the dog alone would be enough to scare off predators. But sometimes the prowling wolf or coyote was disconcerted when the guard dog greeted it and even played with it – after which no self-respecting wolf could steal a lamb from the flock!

Health

Vaccination is the best preventive measure. Most vets recommend a first vaccination for the pup at two months, followed by a permanent vaccination at about 14 weeks and then an annual vaccination.

Talk to the vet about the risks for dog health. The usual concerns are distemper, parvovirus and infectious hepatitis, for which vaccines are available. Fleas and ticks may be controlled by bathing, dusting with flea powder, keeping the kennel clean and possibly by a flea collar ('organic' ones are now available). Internal parasites include roundworm, hook worm, whip worm and tapeworm, for which there are treatments in great number. The main thing is to read the instructions and follow them. Heartworm is an increasing problem in Australia, expecially in valleys. It can be treated with daily or monthly medication to stop heartworm breeding in the dog's bloodstream.

And don't forget, too many dogs are obese: dogs are sometimes killed with kindness!

Tailpiece

Some interesting scientific studies on the relationship between people and dogs have shown that:
- Survival rates a year after a heart attack are higher among pet owners.
- Whereas greeting a person raises blood pressure, talking to a dog lowers it more than resting.

Ducks

Ducks are engaging creatures, and they fit in well if you have a paddock in which livestock is grazed, especially if it has a stretch of water in it. The commercial breeds don't fly, and so your ducks will settle down with the other livestock quite happily. That is, except the Muscovy, which does fly. We had one that took off one day and was never seen again.

A word of warning, though. Ducks are sloppy creatures and their manure is wet, copious and randomly distributed. In small gardens that can be a nuisance, and it would be better to keep hens.

Ducks lay large eggs, with a stronger flavour than hen's eggs and a more porous shell. For this reason they do not keep so well. They will store in the fridge for about ten days, whereas chook's eggs will keep for over three weeks. The larger pores are more likely to admit disease bacteria where ducks are kept in dirty conditions. Still, some people prefer duck to hen eggs. You seldom see duck eggs offered for sale, except in country districts, where they can be found if you look for them.

As with all livestock, you need to be quiet and confident with ducks. Young ducklings should be picked up in the palm of the hand. Older birds may be caught by the neck, but they should be supported with one hand under the body and one lightly on the wings. You should never carry birds by the legs.

Breeds

The *Campbell duck* from England is the champion layer, with 200 eggs a year. They are lightweights, with the drake weighing 2.5 kg and the female 2 kg. They are more upright than the other breeds, and there are actually two colours of Campbell duck: khaki and white. We found the Khaki Campbell easy to rear and very hardy.

The *Indian Runner* is another light, egg-laying breed, slim and agile. For meat production, on the other hand, choose one of the heavier breeds, like the Aylesbury, which is probably the best duck for meat-production.

The *Aylesbury* was bred in the south of England to sell at the fair or market. It is a white bird with a deep body – the males weighing 3.5 kg and the females 3 kg – and it needs a pond to swim in.

A cross between Aylesbury and Pekin gives a duck that is good for both meat and eggs, so it would be a useful backyard bird.

The *Muscovy duck* is an enigmatic creature, which does not quack, and that is good if you have sensitive neighbours. The wild Muscovy was domesticated in

Muscovy ducks make excellent mothers and if you would like to rear ducklings, this may be the breed to choose. Some people keep them just to rear the ducklings of species which do *not* make good mothers.

the rainforests of South America, and was taken to Europe by the Spaniards. It is a rather ugly, large duck: the males weigh 4 kg and the females 3 kg. There are big red growths on the face, called caruncles, rather like a turkey. The colour varies, but most I have met were black-and-white ('magpie').

These birds will lay about 100 eggs a year and since they fly well, they like a high nesting site. They are good mothers and are often used to rear the young of other breeds, but the males can be grumpy and aggressive. The meat is rather dark, but very good. A cross between the Muscovy and other breeds of duck is possible, but the offspring are not fertile.

Whichever breed of duck you keep, you will find that most of them are friendly and interesting birds. They can be alarmed by sudden noises and people, and lights at night can frighten them. Their quarters should therefore be kept away from a road with traffic on it.

Basic needs

The basic needs of ducks are quite simple. They do not actually need a pool in order to thrive, even though they are water-birds, descended from the wild Mallard. Naturally they enjoy water a great deal and find some of their food in a weedy pond. Ducks will much prefer to be given a pool, a dam or a creek, and I think you will enjoy their company more if they are allowed to behave naturally. Commercial duck-keepers, who mainly specialise in ducklings, do not give the birds access to a pond, but they often keep them outside on grass.

Many small-scale duck-keepers create a small, artificial pond. When it gets dirty, it can be drained to a soakaway, cleaned out and refilled with a garden hose. If you enjoy messing about with water, this arrangement might suit you. Cleaning will be essential because a small area of water will get smelly and will harbour disease organisms.

Heavy meat breeds, such as the Aylesbury duck, may not be able to mate except on water, so if you want to breed ducks, a pond will help. If the weather is very hot, and no pond is available, a water sprinkler will help to cool the birds down.

Water-birds without a pool will need plenty of clean water, both for drinking and also to splash over their heads and feathers. Drinkers should be moved frequently, because of the mud that ensues from all this splashing.

Housing

Ducks need shelter from extremes of weather, but they are hardier than hens. Geese are hardier still. All poultry need housing at night to protect them from predators such as foxes and dogs. Young birds can be in danger from hawks.

The duck shed need not be elaborate. It can be a temporary hut made of straw bales that is renewed periodically, or a movable shed on grass. The best arrangement is probably a wooden shed with a concrete floor, which can be covered with clean straw. Space needed will be about one square metre per bird.

Ventilation can be provided by an opening under the roof, covered in wire mesh to keep out wild birds. Along one wall a retaining board can be fitted to keep in an extra layer of straw for a nest in which to lay the eggs. Ducks do not perch, but sleep on the floor at night, so hygiene is important, because the droppings are wet and the floor soon gets dirty.

Eggs are usually laid in the morning, so duck-keepers do not let out the ducks until mid-morning to make sure that the eggs are laid in the house. We had some 'feral' ducks – past tense because the fox got them – which laid eggs in the fish pond: not a good idea either from the duck's point of view or from ours! When you let them out in the morning, the ducks will rush for the door and may injure themselves in a narrow exit. So a fairly wide door is better than the narrow one normally used in chicken houses.

Ducks will spend the day outdoors, but will soon learn to come back to the house at night if you give them an evening feed there. Where will they spend the day? On the pond, they hope. If you have plenty of paddock space, they will enjoy free-range, but keep them out of the garden. If space is short, two grass runs can be used alternately, with perhaps a concrete area for wet-weather use in areas of high rainfall. Another way of keeping ducks is the strawyard system, with plenty of straw, which is renewed when it gets dirty.

Duck pens need to be fenced with fine wire mesh about one metre high. The birds can fly over a fence at this height, but to prevent this, one wing can be clipped. Just cut the tips of the primary feathers to put them off balance for flight.

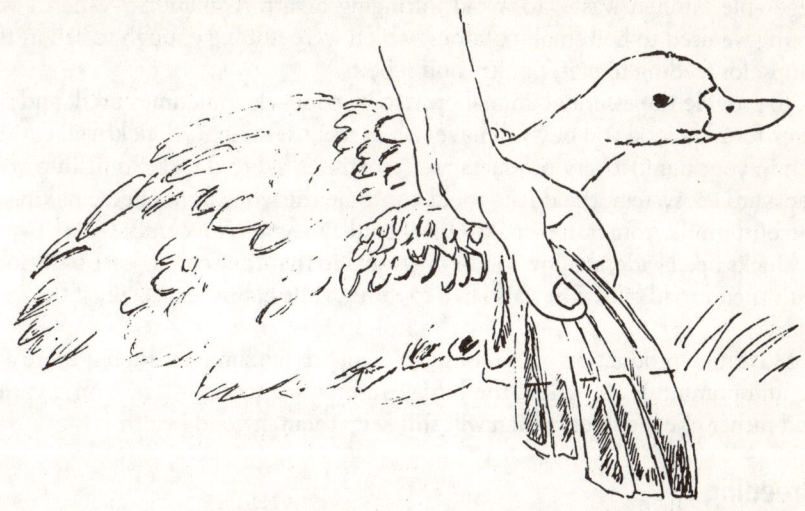

Clipping a duck's wing To prevent ducks from flying over netting and out of the paddock, clip the tips of the primary feathers of one wing. This does not hurt the bird and may keep it out of trouble.

Feeding

Ducks are rather more fussy than hens in what they will eat, possibly because they have better developed senses. They pick up food with the bill, helped by the tongue, and saliva helps it down the gullet, which widens to become a crop. The food passes down in small amounts at a time to the stomach. In the gizzard, as with other fowls, the food is ground by muscular contractions, aided by grit.

Grass and pond weed will be nibbled, and provide minerals and vitamins for the birds, especially Vitamin A. This can be supplied by cod liver oil when there is no green food available. The roughage provided by greens will help the digestion. Very long, coarse-cut grass may be dangerous, because it can form a ball in the stomach and cause a blockage.

Ducks do not graze grass seriously, the way geese do, and so they need a good basic diet of about 150 g of food a day. This can be a mixture of grain, mash, grass and vegetable waste. A mixture of several grains, coarsely ground, is often fed to ducks. Pelleted food formulated for ducks can be bought: starter for the ducklings, then grower and breeder pellets. If you are a nervous beginner, it could be reassuring to give your ducks a complete, balanced feed. But, as with most stock, the food is a major expense, and the aim is usually to watch the cost while keeping up the food value. Most people feed domestic fowl twice a day, usually mash in the morning and grain in the evening.

Mash can be made of stale bread, cooked potato peelings (or potatoes, when they are cheap) and other kitchen scraps. They are dampened, crumbled and dried off with a handful of bran, oats or barley meal. Most poultry and rabbits will appreciate this traditional backyarder's economy feed. It is best to keep to vegetable kitchen waste, to avoid infringing health regulations. When I was young we used to boil small potatoes, which were not big enough to sell in the shops, for feeding to pigs, poultry and rabbits.

To provide the essential animal protein, fish meal is sometimes used, and the duck formula food you buy will have fish or meat meal in it. (Ducks will eat the fish in your dam.) Dairy products are sometimes fed to ducks. To fit into your backyard ecosystem, the ducks could consume the whey from cheesemaking or the buttermilk from butter-making. All poultry seem to love milk.

Ducks need wide, shallow food-containers so that they can scoop up the food with their broad bills. They also like to drink while eating, especially if the food is dry.

Maximum production of eggs or meat demands maximum feed, but there can be an optimum level, where the birds can produce without strain on a varied and rather cheaper diet, which will still keep them in good health.

Breeding

Ducks have courtship rituals, like other species. In the autumn the drakes perform for the females, shaking the tail and displaying their feathers. Attached males swim round their mates, nodding all the time, and flashing the back of his neck at her.

The females display 'inciting' behaviour, swimming towards the male of their choice and making threatening noises to the others. If the drake likes the female, he will drive other drakes away. Bonding gradually takes place and bonded pairs then spend their time together. All the courtship and mating activities happen on the water, which makes me think that if ducks are to live natural lives, water is necessary.

Wild ducks pair and then select a nest site together. The female builds the nest, making a shallow hole in the ground by scraping and then rotating in the depression. The eggs are buried in whatever material is to hand; ducks don't carry material about as other birds do. When the clutch of eggs is complete, the female plucks down from her breast to camouflage the nest.

Like hens, ducks turn the eggs, more frequently as the time of hatching draws near, and the female calls softly so that the chicks know her voice even before they hatch. The ducklings chirp in the shells, unless an enemy comes near, when the female makes a warning sound and the babies in the eggs keep quiet.

The female duck oils her feathers before the babies hatch, so that they pick up the oil and are waterproofed straight away. They don't stay in the nest, but move off with the mother a few hours after hatching. Ducklings need less mothering than chickens, but the mother duck looks after her babies for three or four weeks. The ducklings are not fed by the mother, but are taken to places where there is food, and the ducklings follow her in single file. The baby ducks are more attached to each other than to the mother, and they can manage without her if necessary after the first week or so.

Rearing ducklings

Duck eggs can be incubated and hatched in an incubator, or the job can be given to a broody Muscovy duck or a hen. Domestic ducks have lost the brooding instinct for the most part. About ten eggs can be incubated by the average duck, but their webbed feet can be clumsy, and breakages are a problem sometimes, so many people prefer to use a broody hen. The incubation period for duck eggs is 28 days; or 35 days for the Muscovy.

A quicker way to start with ducks would be to buy day-old ducklings, to be brooded by a hen. She could be allowed to sit on some china eggs for a week, suitably dusted with insecticide. Then, one evening, slide out the eggs and introduce the first duckling. Wait at least half an hour. The next ploy is to scatter food in front of the hen. If she has accepted the duckling, she will encourage it to eat, and then you can introduce a second duckling. With half an hour between each one, this can be a long job!

Of course the hen gets very upset when the ducklings take to water, and in fact it can be dangerous if the ducklings are too young. Ducks have oil on their feathers, which enables them to throw off water. In the nest, the youngsters pick up oil from the mother duck's feathers. But a hen has no such oil, so the baby ducklings are not waterproof.

Feeding ducklings

There is enough food in the yolk sac to keep the newly hatched duckling going for several days, so ducklings are not fed for the first day. For the first six weeks or so, they need seven meals a day, and they are given as much as they will clear up. A special high-protein duckling ration is available, or you can make your own by mixing equal parts of oat flakes and cracked barley with half that amount of bran and finely chopped greens. Mix the lot with water until it is moist and crumbly.

In the absence of feathered broodies, you can brood the ducklings under an infra-red lamp, at $30^{\circ}C$ to start with, decreasing slowly until they can manage without heat at two weeks. The temperature should be checked with a thermometer and is determined by the height of the lamp.

Table ducks can be fattened in six to twelve weeks. Stale bread and grain makes the best fattening ration. The best time to kill them is just before they moult for the first time and acquire adult feathers. Once they grow these feathers, they are very hard to pluck.

Fish

Fish-farming is one of the ancient arts being revived in modern times. A farm dam can provide you with a good source of food, if the right conditions prevail and the right fish are kept.

Dams, of course, have many uses. Extra water-storage is always prudent in Australia, and on many farms the dams provide the only water for stock and crops. They are attractive to wildlife, especially when the surrounding area is planted with native vegetation. Big dams can be used for swimming and boating, and they can enhance the landscape a great deal.

The harvest from your dam will depend on your area. First of all, are you game to try eel-pie for tea, or perhaps eel in tartare sauce? Eels are nutritious and plentiful in many dams, but are not often eaten these days, although Mrs Beeton's famous cookbook suggests seven different ways of cooking them.

Some people buy fish to stock the dam, and this is quite easy, once you know the rules and the regulations. For good results you need a dam at least one-tenth of a hectare in surface area, about 1 000 square metres.

Most fish won't breed in farm dams, so the idea is to buy the young fish (fingerlings) and grow them to table size, restocking every three years or so.

It is not easy to work out how many fish to stock, but the general rule is about 250 Trout or Golden Perch to the hectare, or 350 Silver Perch or Catfish. You can buy the young fish from hatcheries, but wait until you've used up all the older ones before restocking. Otherwise, the older ones will probably eat the new arrivals.

Transport the young fish carefully and follow the hatchery's instructions. Place in the dam the container with the fish in it and they will eventually swim out into their new home.

It is also better if you can keep the other stock out of the dam, or restrict them to one end, to avoid too much fouling of the water. But muddy water is not a problem, and you don't need water flowing into the dam. Dams with trees round them always look so much better than bare ones, and trees are a source of food for fish, because of the insects that hover around and drop from the branches. Water plants are the best cover for fish, to protect them from birds of prey. But too much vegetation will cut down the oxygen, which is absorbed through the water surface. Sunlight is also needed on the water, in some places at least, to help the growth of algae.

Fish varieties

Eels The usual dam with pond weed and the usual stock of aquatic life will probably have eels in it, without any expense of stocking. Eels eat dragonfly larvae, which can attack small fish, so they may be an asset, although some people suspect eels of eating their young fish as well.

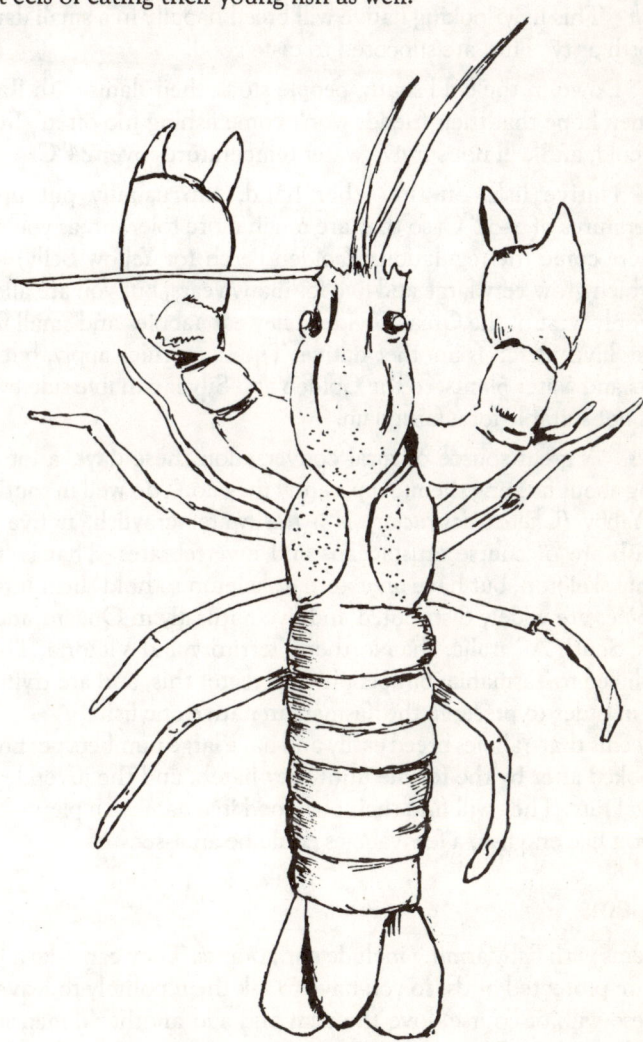

Yabbies can be grown for fish food as well as for human consumption. In some areas it is legal to collect them from creeks or dams, but there may be restrictions on the method of catching them.

To stock your dam with wild yabbies, after checking with the fisheries Inspector, collect them in spring and carry them home in cool damp hessian bags, not in water. The usual stocking rate is 200 per hectare, with equal numbers of males and females.

Eels are quite easy to catch with a line, but they are hard to prepare until you get the knack of it. The skins are tough and hard to remove, and the pioneers cooked them in the skins, so this may be the best way. We cook fillets in the frying-pan; they are a bit oily, but quite pleasant. The old-fashioned eel-pie had lots of herbs and flavourings in it: try Mrs Beeton's recipe.

Catfish This ugly-looking native will breed happily in a small dam, and it may be worth a try. They are supposed to taste good.

Trout Down in the cold south, people stock their dams with Rainbow Trout, and then hope that their friends won't come fishing too often. Trout like water to be cold, and will not survive water temperatures over 24°C.

Perch Native fish, on the other hand, can usually put up with water temperatures of 8–32°C, so they are much more tolerant, as you would expect. But here come the regulations: Golden Perch (or Yellow Belly) are wonderful fish which grow very large and live for many years, but you are allowed to stock them only west of the Great Divide. They eat Yabbies and small fish.

The Silver Perch is another native. The same rules apply, but this one eats insects and water plants, so the Golden and Silver can live side by side. Murray cod is not suitable for a farm dam.

Yabbies A great source of rural conversation these days, a lot of people are talking about farming them. Apparently they don't do well in southern Victoria. The Yabby (*Cherax destructor*) is a freshwater crayfish, native to Australia. Crayfish are of course crustaceans and invertebrates. That is, they have no internal skeleton, but have an external skeleton to hold them together.

Yabbies are widely distributed and live naturally in Queensland, New South Wales, South Australia, the Northern Territory and Victoria. They have been introduced to Tasmania, but people now regret this, and are trying to get rid of them in order to preserve the Tasmanian native crayfish.

It seems that Yabbies breed easily and have large numbers per brood. The eggs are looked after by the female until they hatch, and the juveniles have a good survival rate. They will find their own food in a dam with plenty of aquatic life, so if you like crayfish, a few yabbies could be an asset.

Problems

Problems with fish-farming include cormorants. They can take a lot of fish, but they are protected birds, so you have to ask them politely to leave.

Geese will, of course, love the dam and add another dimension, as well as valuable fertiliser. They are good guard animals and are used by some people to keep cormorants away from the fish.

Wading birds, such as Ibis, will eat only very small fish in the shallow water; wild ducks are no problem, but domestic ducks do eat fish.

Too much plant growth can make the dam unsuitable for fish, and so can predator fish. Algae can use up all the oxygen in the water so that the fish die, but of course some algae is needed for fish food. Algae growth is sometimes encouraged by fertiliser run-off from adjoining land. Some people add alum to the water in spring at the rate of 100 mg per litre: you'll have an interesting time calculating how many litres of water the dam holds!

Some rules

- No English Perch (Redfin) or European Carp allowed in New South Wales.
- Australian Bass only east of the Great Divide.
- Murray Cod, Golden Perch and Silver Perch only west of the Great Divide.

It will be best to check the rules in your state before you stock the dam.

Making a dam

If you have no dam, and there is enough room on your land to put one in, you will probably think about it eventually.

There are two main types of dam on our farm. One is part of a watercourse, which flows through it and out on the other side. One day we would like to build a wall and make a larger dam in this area. The other dam is an excavated depression in the ground into which rain water can flow in the wet season. This kind of dam is easier to make, but may dry out for lack of rain. We have excavated several new ones in the last few years.

The cost of a dam varies with the site, but it will include the hire of a bulldozer and driver for many hours (the current hiring rate is about $80 per hour).

The site should be selected with care and with the help of an expert, who will look at the topography and also the type of soil and whether there is clay under the surface. Clay is needed to seal the bottom, and if you have clay on site it will save the expense of bringing it in.

You may also need permission from the local water authority before constructing a new dam, so the next step should be a check with the authorities as to the proper procedure. And in this case, officials can be very helpful and may be able to give you advice that will save money.

To build a dam that won't leak, you have to remove the topsoil from the area where the wall is to be built, and also from the area where the wall material is to come from. It is moved with a bulldozer or a scraper and set aside to be used later.

If you are lucky enough to have a clay subsoil, the clay is removed from the bottom of the dam and built up to form the wall. It is compacted and sealed into the base clay to prevent leaks and possible wall failure. *This is very important.*

The passage of heavy machinery helps in consolidation, so the machine is driven over and over the clay. One cause of dam failure is a bad seal. In case you need to empty the dam in the future, a length of PVC pipe can be incorporated into the wall. The pipe should have baffles at varying intervals to hold it firmly and to prevent water seeping along the outside of the pipe. A $45°$ bend and an extra piece of pipe attached to the inlet will allow you to siphon water from below the level where the pipe goes through the wall. A control valve, another bend and a short piece of pipe completes it on the outlet side.

Accurate levels are essential for a dam wall and a sight level will be a necessity. The top of the wall must be level for its entire length; and the outlet where the excess water escapes from the dam (not the pipe mentioned above, but an open channel) must allow for the in-flow from a sudden torrential downpour.

Many dams have failed, said an expert I consulted, because their overflows were blocked, had insufficient capacity or were too high. The result is excess water going over the top of the wall and eroding it, which means disaster. We go round our dams in very wet weather, checking that the overflows are working.

The overflow itself should flow gently away from the dam, and then, if possible, enter a watercourse. Rocks or concrete can be used to reduce the speed of the water and thus prevent erosion.

When the main dam wall is finished, the topsoil that was removed earlier can be spread over the top and outer edge of the wall. It is a good idea to spread old hay or straw over the wall and then sow some type of binding grass or plant cuttings. A solid mat of something like couch grass will cover the bare soil and prevent erosion.

While you wait for the rains to fill the dam, enhance the area:
- Fence it off from stock. They trample the edges and foul the water.
- Add a few buckets of water from an established dam to 'seed' the new one with aquatic plants and animals.
- Transplant water plants.
- Plant native trees and shrubs in the fenced-off area.

In a few years, wild waterfowl may nest by your dam, and this will be a great compliment to your work.

Small ponds

Dams for water-storage are as big as you can afford, but ornamental ponds can be a useful addition to the whole system. Frogs are useful in the garden for pest control. The goldfish pond can also be a frog pond, with some waterlilies for effect and some watercress for salads. We used to plant wild watercress in the shallows of little pools.

In fact, the swimming pool could be landscaped, fertilised and stocked with fish, and made accessible to the ducks and geese if you prefer a living pool to a sterile one. Many people living close to a forest area have told me that their pool is mainly a water reservoir for fire-fighting, so why not let the pool be part of the whole design?

The surface area of the pond is the one people mention when they talk of size, and this is because the amount of plant food available is related to the surface area. But fish will need a depth of about two metres to get away from predator birds. Shallow water gets hot in the sun, but the depths are cooler. So the pond really needs deep areas for fish and shallower shelves for the birds and plants, and small ponds can be made safer for fish by submerging hideouts for them, such as a hollow log or an earthenware pipe.

Small areas of water need careful management to ensure that they don't become smelly reservoirs for breeding mosquitoes.

Geese

Geese have been kept by peasants for hundreds of years. In England, before the Enclosures, every villager had a few geese on the common. William Cobbett, who wrote one of the first backyarder's books, *Cottage Ecomony* (1823), says that geese are among the hardiest animals in the world. Goose was a traditional Christmas dinner in days gone by, when cottagers raised much of their own food. The goose has never been taken into intensive production, because it doesn't fit into the factory farming methods. Goose meat is rather greasier than other poultry meat, and this is contrary to the trend towards lean meat.

Turkey has replaced goose as a large bird on the party menu, and you cannot buy goose in a supermarket. In fact, I am not sure whether you would like it: the taste is less bland than mass-produced turkey, and there is quite a lot of fat.

The best way to keep geese is as part of an aquaculture system. The main purpose of the system may be to produce fish or Yabbies for food, but waterfowl like geese or ducks have a part to play. Fowl will produce manure for fertilisation to increase the nutrients in the water, which will help the growth of plants to feed the other organisms.

If you do decide to rear geese for meat, don't forget to save the feathers, which are of value, as the old-fashioned goose-feather quilts are popular again. Sterilise the feathers, which often have mites or other parasites in them, by heating them, in a brown paper bag, in a cool oven or by putting the bag in the freezer.

Why keep geese?

There are, however, quite a few reasons why backyarders still keep geese. These birds are much hardier and easier to rear than turkeys, and cheaper to keep. Goose meat is the cheapest of all to produce – trust our thrifty ancestors to work that out – because they are grazers. Their main food is grass, with a little stale bread or barley meal for fattening.

Geese can withstand extremes of climate, hot or cold, because of the insulating layer of down under their feathers. They prefer to live and sleep outdoors, but it is wise to shut them into a simple house at night to preserve them from foxes. Geese are often kept for grazing under orchard trees in a two-tier husbandry system. In autumn they eat fallen and damaged fruit, which can provide a good part of their diet. The geese help to keep down the grass and weeds round the trees, and control the spread of disease by eating the fallen fruit.

If you have a large stretch of grass (and this is necessary as a fire precaution round many Australian homesteads), geese make good lawn mowers. It makes sense to hand over the chore of cutting grass to birds who can digest it! Grass grazed by geese is very short and fine; the droppings will not be a problem, unless your flock of geese is a really large one.

Geese will sound the alarm when strangers appear, sometimes making threatening runs. They have been known to fight off intruders and can be aggressive in the breeding season, especially the male bird. Like swans, they attack with the bill, and also can deal you a blow with their powerful wings. In fact, geese vary in temperament, and some can be placid and friendly, following you around in single file, appearing on the doorstep in the mornings and waiting for you to come out. But some can bite your legs, and for this reason I would not recommend anyone with small children to keep them. I can still remember how our geese bit my skinny legs when feeding them was one of my duties as a teenager. We try to keep geese and small children apart.

Choosing a breed

There are not very many breeds of geese to choose from, and they are all old and unimproved, because geese have never really been bred for production.

Years ago, we kept white Chinese geese for the Christmas trade on a backyard scale, rearing them on a diet with plenty of variety: stale bread, boiled potatoes and household scraps all mixed up with barley meal. These are the smallest breed, and maybe the most suitable for your backyard. The drawback of Chinese geese is their yellow meat. Roman geese are rather larger, while the old European breeds (Embden and Toulouse) are big, heavy birds, which put on weight easily.

Breeding geese

Over the years, we have kept geese on and off and found that there are several ways to get started. Once we bought some baby goslings and reared them in a grass run, without heat after the first two weeks. (They started off under an infra-red lamp borrowed from the pigs.) They were reared on bread and milk and chick crumbs (the commercial food for baby chicks) and started to eat grass when quite young.

Although adult geese are very hardy, the young ones are susceptible to cold until they get their adult feathers. So, in all but the warmest climates, they will prefer to have a shelter for night and as protection in bad weather.

Another time, we bought four young geese, which were just acquiring their first adult feathers. They had been reared by humans and followed us around for a while. Whenever they saw feet they would follow.

If you would like to breed geese, three females and a gander is a breeding set, and the females should lay about 100 eggs a year between them, in the spring. Put them in a shelter at night for protection from foxes and make a straw nest there to encourage them to lay inside. You can, of course, eat the eggs! One goose egg will equal about three hen eggs. Not all geese will go broody, but many people use artificial incubators for goose eggs, or a broody hen. The hen will be

The Embden gander is the most common breed of goose found in New South Wales. It is docile and very heavy and it breeds well. The female will lay about forty eggs in the spring and summer. Embden geese are often crossed with other, lighter breeds.

able to sit on about five goose eggs. If you leave the eggs, the goose may go broody when she has laid a clutch of eggs, probably in a nest of her own making rather than in the nice safe shed you have provided. In that case, most people make a temporary shelter with straw bales around the sitting goose. But if foxes are not a problem, or if the birds are in a fox-proof enclosure, you can leave her alone.

Autumn is the best time to get a set of breeding geese together, to give them time to settle down for the breeding season. They are often rather selective about the choice of a mate but, once suited, they tend to be faithful. You may find, though, that the set you have got together is not really compatible.

Geese are long-lived, so they can be a good investment. They love a creek or dam, but do not need water to swim in. If you have no dam, give them a large container of water and keep it clean.

Goose health

Geese are healthy, as a rule, but they can get a gizzard worm from infected grass, where birds have been kept on the same area for a long time. Poultry worming doses can be put into the drinking water should this be a problem. Infected birds look thin and spend a lot of time alone.

Goats

Goats for meat

Goat meat is quite pleasant to eat, and there is an increasing demand for it from people whose traditonal food it is. I would find it hard to produce goat meat, because the keeper inevitably gets friendly with her goats.

A good form of integration would be to use a small number of goats for weed control (blackberry and bracken), with meat as the end-product. The specialist dairy breeds are usually too bony for meat production, but a rather sturdier type of goat is being developed in Australia for meat. More information on meat goats can be obtained from their society, Australian Goat Meat Producers, PO Box 117, Dickson, ACT 2602.

Goats for milk

How many litres of milk can you use in a year? A good goat in a commercial flock will give about 1 000 litres in a lactation, which can last for about eighteen months or so. A lactation is the length of time an animal gives milk, from the birth of a young to drying off.

The world record is held by an Australian goat of the Saanen breed, which gave 3 500 litres in one lactation. Your backyard goat may produce 800 litres or so on a moderate level of nutrition.

To many people, goat's milk is attractive because it is not produced intensively. In other words, it is not 'factory farmed'. For many years, it has been used for people suffering from allergies, particularly for those allergic to the protein in cow's milk. Another 'health' factor is the low fat content.

The average composition of goat's milk is: 3.8 per cent fat, 8.6 per cent solids (not fat), and the rest water. The fat and protein are present in smaller particles than they are in cow's milk, which makes them more easily digestible and useful food for people with ulcers, liver diseases and some skin complaints. The milk of the goat is mildly laxative and high in phosphates and Vitamin B.

When our family kept goats, we found that cheese-making was easier with goat's milk because of the small fat globules. Less fat is lost in the process. However, the cream is slow to rise to the surface, and so it is difficult to skim off by hand for butter-making. Many goat-keepers have a small electric separator for removing the cream from the milk.

The idea of producing your own milk, cream, butter, cheese and yoghurt is very tempting and, with some care, it can be done.

However, it concerns me that so many people get into goats, and then get rid of them, because of the problems they encounter. So let us look at the goat scene in the more densely populated areas.

Goats are characters who provoke extreme reactions, and they have been 'scapegoats' for centuries, blamed for creating deserts among other things. This is a warning before you begin, but not a good reason for avoiding them. I think that everyone should have the experience of keeping goats at some time or another.

Your goats could be a threat to your environment, but only if they get out of control. Your goats could help you to maintain a balanced system, because there is a place for goats on the smallholding when you produce your own milk or fibre.

Two dairy goats make a very good contribution to backyard self-sufficiency. As we have seen, they can provide enough milk for a family all the year round, and very often there will be surplus milk with which to make yoghurt, cheese or butter. Spare milk can be fed to other animals, such as pigs and poultry. There is nothing like milk for fattening pigs.

Meanwhile, the goats will eat surplus garden vegetables, weeds and bakery waste, efficiently converting all this into milk and manure. They will be lovable pets, and can be taken for walks by the children.

Goats for fibre

Angora goats

Angora goats produce mohair. After a 'boom' period when animals fetched thousands of dollars each, Angora goats are now an alternative enterprise for people who like goats. Breeders say that Angora goats 'complement' sheep and cattle grazing patterns. That means that they have different requirements from the other classes of stock, and can be run with other types of stock to great advantage.

Compared with the wily dairy goat, the Angora is a docile little animal, more like a sheep in habit. In general, Angoras need no more than conventional sheep fencing, not as high as for goats, which is good. The quality of the stock is important if you want good fibre. Pure-bred goats produce much more good quality fibre than cross-bred ones. Two crops of mohair a year are usual, in early spring and early autumn. A pure-bred doe should give you 4–5 kg a year of mohair, and a kid will produce half this amount in its first year.

Five does are grazed to the acre on good quality grazing, but on an acre or two with sheep or cattle, a couple of Angora goats could probably be introduced as well. All types of goat are good at biological weed control since they love to browse on tougher plants.

Care of Angora goats When in full fleece, Angora goats need a little more care. They can easily get caught in scrub if you have a bush block, and although this sort of country is ideal for goats, you will need to check them more often than usual to make sure they are not caught by the fleece in the undergrowth.

Like all goats, Angoras dislike rain, and they should be given the option of shelter, but don't worry about shutting them in a shed against their will. Rainwater is good for the fleece before shearing. After shearing, they will feel the cold and wet for a while, so they should always have the option of a shed. At kidding time you will need to keep a close watch on all goats. Apart from this, Angoras are usually healthy animals with few problems.

Cashmere goats

These goats are bred from feral goats producing a very fine downy undercoat, the value of which was first realised in Australia in 1972. The fibre is short and fine, ideal for luxury knitwear. This undercoat is found on goats that do not produce mohair; and it grows as a winter protection when the day length shortens. There is a coarse outer coat, which is present all the year round, and which protects the cashmere underneath.

Cashmere was traditionally produced in places like China, Mongolia and India. There it is a third crop, after meat and milk, from the herds of goats, and it is combed out from the coat. It is one of the finest and most expensive fibres in the world. For the last ten years or so, feral goats have been farmed for cashmere in Australia, and here they are shorn rather than combed, which means that the fleece has more lustre than when it is combed out as dead hair. The aim is to shear in late winter, just before the coat would be shed.

Cashmere production involves processing machinery to separate the fine hair from the coarse, so cashmere has not been a backyard product, but one for farmers seeking an alternative enterprise. However, some people hand-comb their goats and sell the combed fleece to a craft shop.

Choosing a breed

There are several breeds of goat to choose from, some better as milking goats, and some better for the fibre. The dairy breeds vary quite a bit in size and temperament, and the breed you choose should fit your conditions.

For small farms and large gardens the best breed is probably the Saanen, which is the dairy goat most often kept in Australia. It is a docile, cream-coloured animal, quiet and happy to be tethered. A more active breed is the British Alpine, which is a good goat if you have access to bushland or want to clear areas of blackberries.

For the northern states of Australia the Anglo-Nubian may be best, since this goat with floppy ears will tolerate extremes of temperature. The Nubians I have known are very vocal and tend to be excitable. When you start to talk to goat-breeders, you will find them very attached to their animals, even more so than most other keepers of livestock. All of them will tell you that whatever breed they keep is the best.

With all animals, the 'strain' is more important than the breed. In other words, it is important to buy a good type of whatever breed you have chosen. This means going to a reputable breeder to buy your goats. You may drive through a country area and see a goat farm with a 'kids for sale' sign, but before you go through the gate and fall under their spell, do some homework and find out who they are and what their reputation is.

Agricultural shows are the best places to start. Get into conversation with goat-breeders at shows: this is why they are there. Look carefully at prize-winning animals so you will know what a good example looks like. Once you start to look for goats, you will find them everywhere, often tethered, grazing on roadsides, the banks of dams and any weedy areas.

Read whatever information you can come by, and, once again, check the regulations. In the country town I looked at (see page 5), you may keep no goats in the central business area, two in the suburbs, and as many as you like on the outskirts of the town.

Having read and talked about goats, the next stage is deciding on what suits you. You can buy weaned kids, goats in kid or in milk, or young females to rear. A kid will need milk for the first six months or so. A goat in milk will give you an instant return, but you will need a few lessons in milking before she arrives. A young goat, in kid, will get time to settle down and get to know you before she gives birth and milking begins. If you are nervous about kidding, it may be best to get a goat in milk so you won't have to worry about another generation for a while. Many goats will milk for two years after kidding, without drying up.

Feeding

Hay, pasture grass, grain concentrates and branches of trees and shrubs make up the usual balanced goat ration. Goats need a high proportion of roughage, with plenty of 'browse' or branches, but milking goats are usually fed a concentrate ration as well, usually at milking time. This has the advantage of keeping them quiet while milking is proceeding, and ensures that they come willingly to be milked.

The table shows the recommended *daily* quantities of feed for goats. These are average figures. The actual amount needed depends on the size of the animal, her body condition, and on the amount of milk she is producing.

Food	Kids (1-6 months)	Kids (6-12 months)	Goatlings	Milkers
Milk	2.25 L	–	–	–
Concentrates	up to 250 g	250-750 g	250 g and grazing 500 g without grazing increased during pregnancy to 1.5 kg at kidding	500 g - 1 kg per 10 L milk
Hay		all ages, as much as they like		at least 2 kg/day
Roots	a little	1 kg	2 kg	3 kg
Grass and grazing	handful cut grass or grazing	1 kg cut grass or grazing	2 kg cut grass or grazing	10 kg cut grass or grazing
Minerals and vitamins	all ages, high phosphate mineral supplements and cobalt lick			vitamin C before kidding

A cow concentrate ration can be used for the grain part. Some people feed whole grain, while others mix up a special goat ration of their favourite recipe. One example is a muesli mixture of equal parts of bran, crushed oats, flaked maize and linseed cake. Goats are queer creatures, and quite fastidious when it comes to food. They will drink only really clean water, and will not eat food that is in any way soiled or stale. So food and water troughs need to be kept clean; and any stale food, or any that has fallen to the floor, removed.

Housing

Goats are also rather strange about housing. They are incredibly hardy, living in high mountain districts, but they hate wind and rain. You will find that tethered goats expect to be moved in if there is a shower of rain, and they really need draught-free shelters and warm, dry beds. An existing shed can be made into a goat shelter: separate pens for each goat is a good idea, with strong wire mesh partitions so that they can see each other. About three square metres will give enough room for two or three animals, and a yard for exercise should be larger, say three metres by four.

The easiest way to keep goats is to give them a shed and yard, so that they do not need to go out on the land every day. Tether them when there is good food for them to eat, some weeds to clear, and you have time to look after them. It is not wise to tether a goat and leave it alone while you go out to work for the day; but a yard system allows them fresh air while keeping them out of trouble. The pen should have a large feeder to keep hay and green food off the floor.

Goat problems

The problems with goats arise mainly because they are browsers: they like to nibble at bushes, trees, and flowers – anything for a change. Although small, they are extremely agile and quite intelligent, so that they need very good fencing, unless they are tethered. A goat with too much freedom would probably ruin your tree-planting programme, then your garden, and finally your friendship with the neighbours.

The other objection raised by goat-haters is that goats smell. That is true of male goats, but not of well-kept females, and when the milk is produced properly with good attention to hygiene, it does not have a goaty odour either. Most backyarders do not keep male goats: they prefer to travel to a stud with the females when mating is due.

Goat care and health

Cleanliness is important for all livestock health, and sheds should be kept clean and as free from flies as possible. Parasites can be a real problem for goats. Like the other grazers, they can be infected with nematodes, tapeworms, lungworms and liver fluke. These can cause misery to the animals and loss of money to the goat-keeper. Vets say that a lot of people lose money by:

- Worming at the wrong time.
- Using unsuitable products.
- Not estimating the weight of the animal correctly and therefore underdosing.
- Losing milk where the poor things have a heavy worm burden.

Only good management will reduce the effect of any kind of parasites. Grazing paddocks should be used in rotation as a precaution against parasitic worms, since the eggs that fall onto the pastures in the animal droppings will die after a few weeks if they are not picked up by a goat again. The bacteria that cause foot rot will likewise die out after a few weeks. Three weeks is a good rest period for a paddock in the growing season.

If possible, paddocks should be rotated from year to year so that you grow a crop and then reseed with grass. Try to keep clean pastures for kids; that is, areas of grass that have not been grazed by goats for a year.

The other important point is to ensure that the goats get all the minerals and vitamins they need, since healthy goats are so much more resistant to parasites. Liver fluke is sometimes a problem in wet areas, but ducks are the organic solution. Ducks will eat snails which are the alternate host of the liver fluke. So if you live in a 'flukey' area, include some ducks in the system.

Most commercial goat-keepers drench the stock for worms (and, in some areas, for fluke) at kidding. You may wish to avoid the strong chemical drenches, and use herbal treatments such as garlic instead. In that case, extreme vigilance is needed to keep the goats free from internal parasites. Watch, particularly, newly kidded goats, kids and debilitated animals, because they are the most susceptible to parasites.

Externally, goats may suffer from lice and can be dipped in sheep dip. Backyarders can keep an eye out for parasites and give their goats a dip if necessary. In the old days, a mixture of soft soap and carbolic acid was used.

Goats are normally healthy animals. By watching their behaviour, you should be quick to spot any ailments. Danger signs include loss of appetite, isolation, no ruminating, drop in milk-production, scouring. Call the vet sooner rather than later when you are learning, because a good vet will give you much valuable advice.

Mastitis

A hot or swollen udder, perhaps with clots in the milk, probably means mastitis. The usual treatment for mastitis is an antibiotic intra-mammary infusion, but you will need a vet's prescription for this, so seek advice anyway.

No doubt you will want to use antibiotics as sparingly as possible: they have their place in coping with serious outbreaks of disease, but should be kept for emergencies. A mild mastitis will need watching, as it will be very painful for the goat and may get worse. Frequent milking out, so that there is no milk left in the affected udder, helps to get rid of the infection, and hot compresses seem to help too.

Since mastitis is very easily transmitted from one animal to another, keep an infected animal apart, and milk it last if you have more than one milker. Don't drink the milk, and check the withholding period for the milk if you do use an antibiotic. Apart from the fact that you don't want a dose of antibiotic, the treated milk will be useless for dairy products, because the antibiotics will kill the bacteria you need to make yoghurt or cheese.

Hoof trimming

Hoof trimming is another regular job, unless your goats can wear down their hooves naturally on rocky ground. The hoof is trimmed back level with the sole of the foot, using a sharp knife. Do it for the first time under the guidance of an experienced goat handler. Groom the goats frequently: a good brush before milking may remove loose hairs and dust.

You don't throw a goat over to trim the feet, but leave her standing like a horse or cow, safely anchored to a post. Tie the goat up short, against a wall if possible, and let her see what is happening. Make sure you have control before you cut.

The diagrams show the stages in trimming a badly overgrown hoof. This foot has been neglected, germs could get in, and the goat could be lamed. The footrot shears are used to pare the excess hoof, trimming away excess nail, and then paring the heel so that the foot is level. The shears can be used gently to dig out dirt from between the toes. The sharp knife is used to finish off by thinly paring the soft tissue.

Trim several times if the hoof is long, rather than doing it all at once.

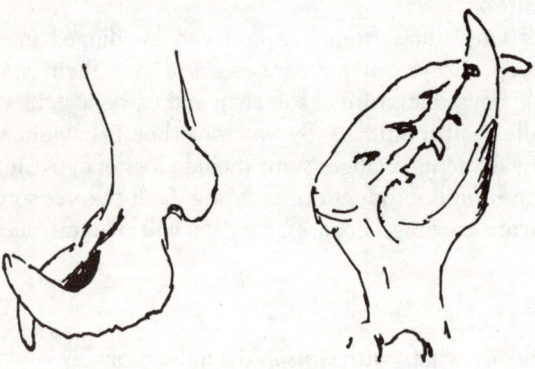

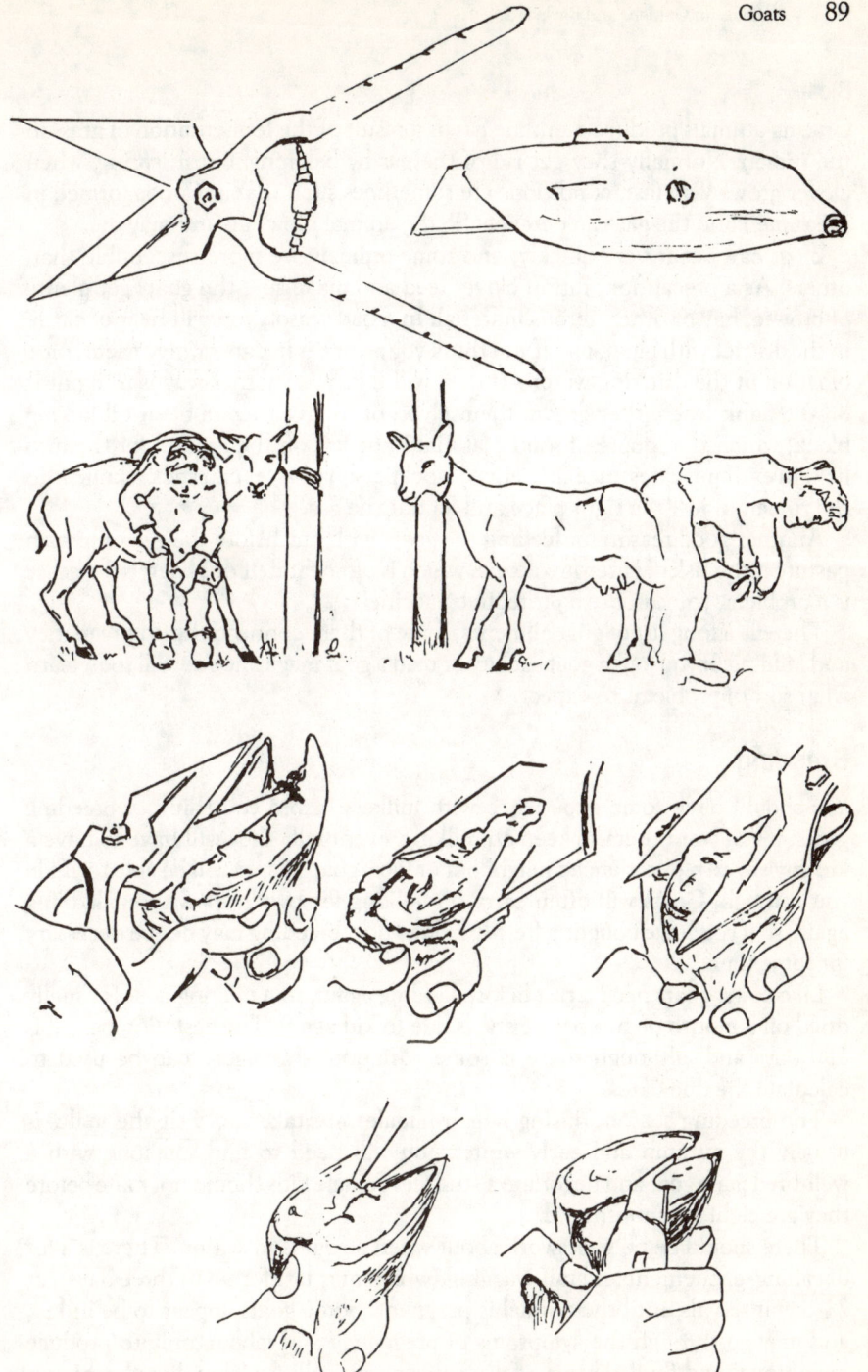

Bloat

Grazing animals produce methane gas as a result of the fermentation of grass in the rumen. Normally they get rid of the gas by belching, but in spring, when clover grows very fast, conditions are sometimes such that a froth is formed in the rumen and the gas can't escape. So the animal blows up and may die.

Bloat can occur very quickly, and some animals are more susceptible than others. As a precaution, ration clover feed and make sure the goats get plenty of browse, hay or other fibrous material. In a bad season you will hear of cattle in the district with bloat, and if you think your stock is in danger, give them some bloat oil in the drinking water – if they will take it! Another way is to paint it on the flank where they groom themselves, or to give them a bloat-oil licking block, which is a square of solid material to be left on the ground for them to lick. In extreme cases, an experienced stockperson or a vet can stick a knife into the rumen in just the right place and let out the gas.

Another good reason for feeding the goats with hay before they go onto lush pasture is the risk of enterotoxaemia, which is often fatal. If this bacterial disease is a problem, you can use a protective vaccine.

There is a long list of goat ailments, some of them connected with pregnancy and kidding. If you join a goat society or read a goat magazine you will soon learn what sort of problems to expect.

Breeding

You should have some experience with milkers before you think of breeding goats. If you want them to keep on milking, eventually they will have to have a kid anyway. An experienced neighbour or local goat-keeper will be able to guide you through. Goats will often carry on milking for two years without kidding again, so if you have bought a freshly kidded doe, breeding may not be necessary for some time.

Like cows, goats need a rest before kidding again, so a milking goat is usually dried off a month or two before she is due to kid again. The gestation period is 150 days and, although there is some variation, this figure can be used to calculate the due date.

The breeding season, during which females are taken to visit the male, is usually the autumn and early winter. You will need to find someone with a well-bred male goat and negotiate a stud fee. Female kids should not mate before they are eighteen months old.

There should be no guesswork about when a goat is in season. There is a lot of calling, excitement and tail-wagging (which may last for up to three days), at 21-day intervals until the animal is pregnant. Some goats appear to be in kid, and may go through the symptoms of pregnancy and labour, only to produce membranes and fluid. This is a false conception, called a 'cloudburst', which is most disappointing both for you and the goat.

When goats kid, don't panic. I have come to the conclusion that when animals give birth, it is best to leave them alone. Be aware of what is happening and be ready to help or to get help if they are in trouble. But there is no need to interfere with a normal birth.

As with other animals, a goat about to give birth will be restless. The udder will fill and the pelvic bones loosen. Don't leave buckets of water in the pen, as the kid may be dropped into one. Normal presentation for the kid, as in sheep and cows, is with the forefeet first, hooves facing down and the head resting on the legs, but breech presentations are quite usual.

The goat should lick the new kid dry, but you may like to make sure the mouth and nostrils are clear so that it can breathe. If the goat doesn't lick it, rub the new arrival dry with an old clean towel. There may be other kids, goats often have multiple births. Then will come the placenta.

Young kids, especially Angora kids, may need their bottoms wiped for a few days to prevent unwelcome attention from flies, since their first excretion is yellow and very sticky.

Don't feed the newly kidded goat too much food at first. The milk supply will gradually increase to meet the needs of the growing kid, and this is when the mother will need more food. If your main aim is milk production, kids can be bottle-fed after you have milked the goat. Feed at blood heat three times a day.

Bottle-feeding is quite common for kids after the first few days. They get four feeds a day for the first month or six weeks, then three feeds a day, gradually decreasing until they are weaned at about six months old, when the kid is eating solid food. The amount is gradually increased to a peak of about 2.5 L per day for each kid. If you have milk to spare, whole milk is the best food, but you could mix it with milk substitute as sold for calves.

Hens

Wild ancestors

If we go back to the wild ancestor of the domestic chook, we find a jungle fowl scratching about on the forest floor in warm conditions, eating seeds, insects and green shoots. It loves young green plants, but can't digest cellulose. (This is the tough stuff in plant cells that we can't digest either.) The fowls roost in the branches of trees at night, a closely knit group of a few hens protected by a colourful and aggressive cockerel. The hens nest in private, deep in cool, dark undergrowth, which is rather moist. An egg is laid every day for about twelve days, and then the hen sits tight on the nest, brooding the eggs. They are kept warm by her body heat, so that the embryo chicks develop. While they are still in the shell, the hen talks to them and they get to know the sound of her voice. The shells are kept moist by the undergrowth, so that they will be just soft enough for the chicks to be able to crack them.

After 21 days the chicks hatch, struggling out of the shell with enough food inside them for one or two days, just long enough for them to learn to peck for themselves. The hen fusses over them, protects them from enemies and will keep them warm under her feathers for several weeks until their own feathers are fully grown. She will not actually feed them, but will scratch up food for them and encourage them to eat by making clucking noises.

Domestication has changed the jungle fowl in several ways, and some of the changes are quite recent. Broodiness, the essential instinct to sit on a clutch of eggs and perpetuate the species, has been discouraged by poultry-breeders because it interferes with maximum egg-production. So you will find that the hybrids or cross-breeds, the birds bred for large units over the last 20 or 30 years, just go on laying for a year and they rarely go broody. Our traditional breeds may go broody in spring, but not all of them will want to raise a brood.

Another change is that we have tamed our fowls to a large extent. They are not so wary as wild birds, and although sudden movements can still startle them, most fowls will settle down quite happily in a small area with other animals about. (Guinea Fowl retain more of the wild bird's flight instinct.)

Numbers and cost

It can be exciting to start up a small poultry unit in your backyard. About a dozen hens will keep the average family in eggs for most of the year, and there are ways of preserving a surplus of eggs to cover the flock's vacation period (they give themselves a break every year). Or you can use eggs in a barter system, as many people do, swapping your surplus for something that you don't grow. We are allowed to keep up to 20 birds in a backyard flock in Australia. Over that number you are classed as a commercial producer, which means complications, such as licences and quotas. The big expense is food, and many people will tell you that the food is so expensive that keeping hens is not worthwhile. Each bird needs about 120 g of fairly high protein food a day and the challenge is to provide this economically. Birds that can forage 'free range' will pick up insects and green food for part of their diet. They are seed eaters, and seed can be grown for them, the variety depending on your climate. Sweet corn and tagasaste are two possibilities. In a small backyard, you may have to forage for waste food for poultry. Can you find enough kitchen scraps and bakery waste to keep them fit and laying eggs?

Choosing a breed

While the hybrid layers were developed as egg-laying machines, other crosses were bred for meat, so poultry keeping became more specialised. Supermarket chickens are usually a hybrid from the White Rock and Cornish Game breeds. A bird weighing 2–3 kg will take only 50 days to fatten on a special-formula diet.

To obtain eggs and meat with real flavour, backyard producers usually stick to the older, 'unimproved' breeds and a more natural system. Of course, these old breeds of poultry can still be found, and they are much more interesting than the hybrids. Many of the old breeds were dual purpose, used for eggs and meat. Naturally they will eat more than the economical crosses, and sometimes they will go broody, so you can let them rear chicks in the natural way.

The Rhode Island Red is one of the world's workers, a compact and docile dark red hen, which lays plenty of brown eggs. Since childhood I have liked 'Rhodies', and they are one of the best breeds for the home producer. If you prefer white eggs, choose the White Leghorn instead. A flock with white hens and a red cockerel will breed chicks in which the colour is linked to the sex of the chick: you get white boys and red girls, so you can tell them apart from birth. (The other way round, with red hens, doesn't work.)

For hardiness and economy the best fowl of all is the Bantam, a miniature bird in several breeds, which lays smaller eggs. The Bantam is relatively 'unimproved' by commercial interests, because the eggs are a bit too small. So Bantams have continued as nature made them, living long and busy lives, keeping out of factory farming and laying eggs for backyarders.

In spite of domestication, fowls still like the conditions they enjoyed in the wild. They like to be able to scratch around for food. They like to have a dust bath to rid their feathers of parasites. They need clean water to drink, but they don't like getting their feathers wet. Fowls hate cold and wet weather, so in most climates they appreciate a hut to sleep in at night and to retire to when the weather is bad. The perching instinct is still strong, so they need perches to sleep on; and the door should be shut at night to keep out foxes. Hens still like to lay their eggs in a dark and private place, so they need nest boxes. Good poultry-keeping systems will give the birds as much freedom as is practical, depending on how much space there is, and will allow them all their little luxuries. (The battery-cage system, where a hen can't even stretch her wings, doesn't bear thinking about!)

Breeding

You will have gathered that the question is: 'Which comes first, the chicken or the egg?' If you've never kept chooks before, the most foolproof way to begin is to buy point-of-lay pullets of about five months old. These young ladies will be at the start of their laying career, a most interesting time.

Point-of-lay pullets will be the most expensive way of starting, but you will have some return almost immediately, as they go into lay. If you have chosen bantams or one of the old breeds, you will be able to breed from them later and rear your own replacements.

When you buy them in, feed them the food they are used to for a while, to minimise the shock of change. Get to know them, then introduce changes in the diet gradually. If you feel adventurous or economical, you could buy fertile eggs from a flock with roosters. It does not matter whether the eggs have been kept warm, provided they are not more than a week or so old. Development of the chick embryo stops when the egg enters a cold environment and starts again when the egg is incubated. This is so that a hen can lay her dozen or so eggs over as many days. They develop together when she sits on them to keep them warm. Then, miraculously, they all hatch at the same time. These eggs may cost you a bit extra if the flock they come from is a special one; or they may come free from a fellow backyarder.

Breeding from eggs

The next thing is incubation. You will probably not wish to emulate a man I once knew who hatched eggs in his shirt. The cheapest way is, of course, the natural one: a broody hen. There is a traditional trade in broody hens in country districts. A broody hen is a nuisance to her owner unless she has eggs to sit on, so you may be able to borrow one.

When a hen is broody, she will sit tight on the nest all day, talking in a particular, broody way that you will get to know. She can handle twelve or thirteen eggs (fewer if you are asking her to sit duck eggs), and should be segregated for the duration in a coop with a detachable front.

Cut a large turf to cover the floor and put it in the pen upside down. This will provide the necessary moisture. Scoop a hollow in the turf and line it with hay or straw. Try the hen on the nest for a day, and if she sits tight, introduce the eggs quietly and unobtrusively, at night. Leave her with whole grain, fresh water and grit. Give her peace and quiet and all should be well.

After seven days, you 'candle' the eggs. When the egg is held against a bright light (it used to be a candle, hence the name), the egg's embryo chick will cast a shadow. If there is no shadow, the egg is infertile. Or you may prefer to wait

A hen of one of the old breeds will 'go broody' in her own time, and rear a batch of chicks.

until hatching day, to see what luck you have had. The old-timers check at seven days because they never waste anything, and they know that after a week the infertile eggs can still be used for other purposes!

In hot weather, the last week of the three may be rather dry for the eggs, so they can be sprinkled with water. Watch out for chicks by the twentieth day. If some emerge before the others, it may be wise to take them away and keep them warm until the whole lot are hatched. Watch out for the hen: she may be protective and peck you.

Hen and chicks can live in a coop with the front removed, giving them access to fresh grass. Give them a chick drinking fountain and make sure that they can all drink. You can buy chick crumbs for the babies, a special ration formulated for young chicks, or mix them a ration of breadcrumbs and chopped greens with small bits of kitchen waste.

For the first few weeks, chicks usually stay under the hen's wings at night. They will probably need mothering for about seven weeks, until they are fully feathered and independent. This process can be short-circuited by buying day-old chicks and giving them to a broody hen to rear. She can sit on dummy eggs for about a week, after which the chicks can be put under her so that she thinks they are her own.

Artificial brooding

Brooding is the stage after hatching, when the chicks need warmth and security. Commercial producers use incubators for hatching eggs, then rear the chicks in artificial brooders; this stage is quite easy for a small producer. Chicks need a temperature of about $20°C$, without draughts. To achieve this, a simple brooder can be made with a lamp and a wooden or cardboard shield (to stop them from straying too far). The temperature ranges for baby chicks should be as follows:

Age (days)	Temperature (degrees C)
1–7	32–35
8–14	29–32
15–21	26–29
22–28	23–26
29–35	21–23
36+	21

You can easily tell whether the chicks are comfortable by their distribution in the area. If they are too cold, they will huddle under the lamp, and if they are too hot they will stay at the outer edges. Baby chicks need food and water in shallow trays from the first day of life.

Poultry keeping systems

Free range

In this system the hens usually have a shed in which they can be shut up for the night, with perches for sleeping and nest boxes for egg-laying. They have free access to the open air and to the land, with a density of not more than one hen per 10 square metres, and with the ground mainly covered with vegetation.

This is the usual definition for people who officially describe their eggs as 'free range'. In practice, it means that you give them a fair area to roam in. The presence of vegetation usually indicates that the land is not overstocked. If any animals are kept for too long on a limited area of land, the land becomes sick, in this case 'fowl sick'. It will be bare, becoming a dust bowl in dry weather and a mud bath in the wet. Erosion is then a threat. Fowl-sick land will carry disease organisms to a point that could affect the health of the flock. Our chooks roam free on the range, but we are thinking about giving them a big run with definite limits because they prefer the garden to the paddock.

Advantages of a free range
- The hens have freedom to behave naturally.
- They have a varied environment.
- They have grazing and diet variation.
- You get better, tastier eggs, with deeper-coloured yolks and more vitamins.

Disadvantages of free range
- You need more land.
- Hens eat more food because they use more energy.
- Eggs may get dirty in wet weather.
- Egg-production may be lower.
- Predators can be a problem.

Deep litter system

In this system the hens have the free run of an airy shed with dry litter on the floor in which they can scratch about.

There should be not more than seven hens per square metre of floor space, according to the usual codes of practice.

Advantages of deep litter over free range
- Predators are not a problem.
- Chooks are under control and use less space.

Disadvantages
- Housing is bigger, more expensive, and needs insulation.
- Litter needs attention to keep it dry and friable.
- Chooks depend entirely on you for their food.

Strawyard system

The hens have a roofed but otherwise open-air run, which is covered with straw to keep it dry. They also have a shed for sleeping and laying. Less land is needed for this system, and the birds can choose whether to be inside or outside. But the straw is hard to keep dry in wet weather, even with a roof.

Alternate runs system

The shed stands at the junction of several enclosures, which can be used in turn. It gives the hens less space than free range, but when the run needs a rest they can be moved to another run.

Combined systems

Sometimes two of these systems can be combined. For example, in areas where winters are wet, you might prefer to let the hens out in dry weather and to keep them in the shed on deep litter when it is wet. Your choice of housing may also depend on what other stock you keep, and how much space is available. On small farms and in gardens it might be downright inconvenient to have the chooks roaming about all through the year. They may be welcome in the garden at certain times, to pick off garden pests when the plants are well grown, or to wander over a paddock after you have cut the hay.

With these considerations in mind, it is a good idea to make a shed big enough for your flock to live in a self-contained deep-litter system when needed. From there they can be let out or not, as you decide and as conditions dictate. Whatever system you use, the hens will need access to food, water and nest boxes, protection from predators and insulation from extremes of temperature. They are prone to heat stress and will need extra water in very hot weather. High humidity with high temperature causes more stress, as it reduces their capacity to lose heat by evaporation.

Housing

Don't start to build with my suggestions in one hand and a hammer in the other. Check the local regulations first. The shed should be strong enough to withstand a gale and insulated enough to keep out extremes of heat and cold.

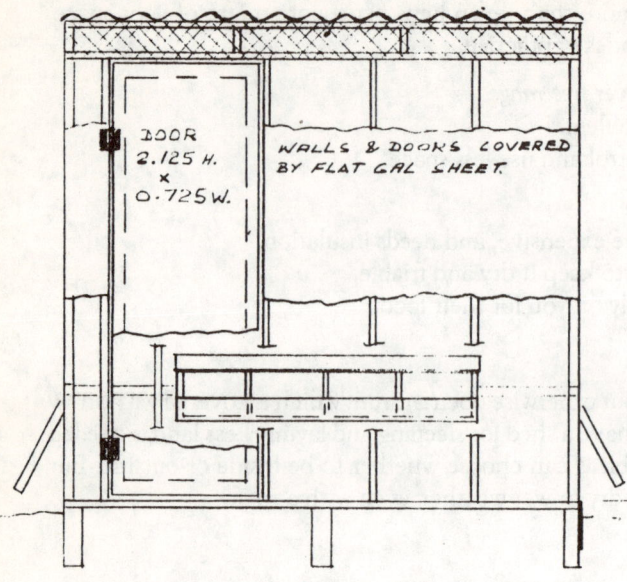

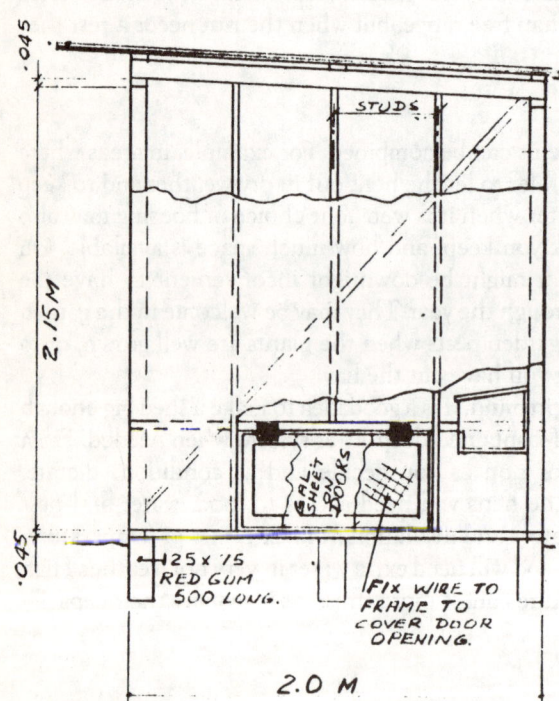

Common specifications
- Red gum stumps.
- Frame of 75 x 45 mm hardwood.
- Studs at 500 mm centres.
- Roof rafters 100 x 38 mm on edge.
- Noggins 0.5 m above ground level to allow for air flow in summer; the opening to be covered in winter with hinged flaps.
- Door in front about 0.75 m wide.
- Flat galvanised-iron, taken 0.5 m below ground level, and fixed to noggins and bottom plate all round.
- Gaps between rafters and ventilation gap to be covered with 12 mm wire mesh to keep out wild birds.
- Cladding: iron sheeting, corrugated roof nailed to the purlins on rafters.
- Bracing: frame braced with 50 x 25 mm timber to keep it square.
- Floor: 50 mm concrete will help to keep it clean.
- Perches: timber 50 x 25 mm laid flat, supported at intervals by heavier rails, suspended by wire from rafters or nailed to studs 0.6 m above litter; 300 mm apart and allowing 250 mm per bird.
- Nest boxes: at least one per three hens; old wooden fruit boxes suitable; measurements: 230 mm wide x 350 mm deep, 280 mm high with a front plate about 100 mm high to keep the straw or wood shavings or rice hulls in the box.
- Sloping roof, boxes fixed about 0.4 m above the floor.

Size
Half a square metre per bird is the minimum. You could be more generous if they have to spend much time indoors because of local weather conditions.

Design
Look around and you will see many chookhouse designs, some of them extremely original. A square shed or lean-to with a sloping, skillion roof is common. The front of the roof can be extended a little way out.

Litter

I have used straw for deep litter, but it tends to get soggy and then pack down. Rice hulls are what we use now. You can buy a small bale of rice hulls and cover the floor with a material that hens like to scratch about in. Wood shavings and sawdust are other materials used for deep litter, but they break down more slowly in the soil when the litter goes onto the garden.

Good deep litter is about 200 mm deep. In it, bacteria break down the droppings quickly, and you have a dry compost material with a faint smell, not enough to attract flies, which is important in a garden setting. Poultry manure is a valuable garden fertiliser, so it is worth a little trouble to get the litter working properly. If it goes wet and smelly, add garden lime, turn it over and top up with dry material.

Feeders and drinkers

Any sort of shallow trough will be useful as a feeder, and a trough will be needed if you feed the poultry on scraps and mash. Make sure there is enough space for all the birds to feed together. Galvanised troughs are available, and they are easier to clean than the old-fashioned wooden ones.

If you are feeding layer pellets, the best poultry feeder is the hanging tube type, which can contain enough hen food for a couple of weeks if you use the conventional pellets. With this type of feeder, the food is available all the time, so it doesn't matter if the hens can't all feed at once. There is a small space at the bottom of the tube for the feed to fall through. The feed hopper is suspended from a rafter and hangs about level with the birds' beaks. If you feed grain to the birds, scatter it in the litter: they love to forage for it.

A large shallow container can be used for water, but it will need frequent cleaning, as hens are messy birds. The fountain type of drinker is better, or you can buy an automatic drinker, which is the cleanest type available. Poultry need plenty of good clean water at all times, about 250 mL per day, and much more in hot weather. All birds need water to soften the food and help digestion.

People with small gardens can confine the chooks in a roomy shed, hanging up bunches of green waste from the garden for them to peck at. Weeds, outside leaves of cabbage and lettuce, and grass cuttings will all give the hens an interest and make the egg-yolks dark-orange with carotene.

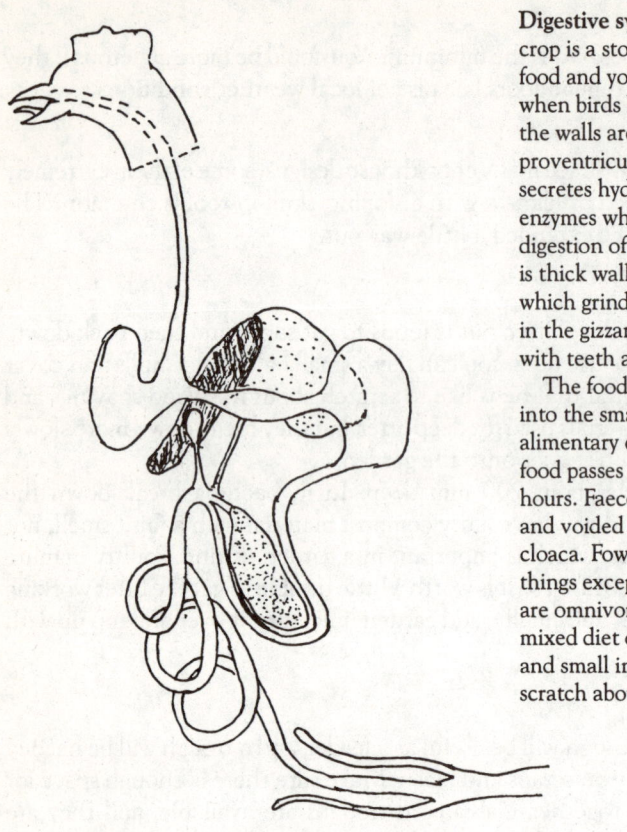

Digestive system of a hen The crop is a storage compartment for food and you can see it bulging when birds have eaten because the walls are thin. The proventriculus is a gland which secretes hydrochloric acid and enzymes which are needed for the digestion of the food. The gizzard is thick walled, with muscles which grind the food with the grit in the gizzard (instead of grinding with teeth as we do.)

The food passes after grinding into the small intestine and as the alimentary canal is quite short, food passes through it in a few hours. Faeces and urine are mixed and voided as droppings from the cloaca. Fowls can digest most things except cellulose and they are omnivorous. They like best a mixed diet of grain, green shoots and small insects and they like to scratch about to find them.

Fowl biology

Let us take a look at the biology of the bird, which helps us to understand our hens and to know what to expect of them. A female chick, either brooded by the hen or hatched in an incubator and reared in an artificial brooder, will grow rapidly for the first few weeks and then more slowly. At about 18 weeks of age she is mature and will be transferred to laying quarters. The actual age at maturity depends on daylight length, as do so many events in the breeding cycles of animals and birds. The more hours of daylight a young hen sees, the more quickly she will mature: the range is about 17–24 weeks.

Artificial light is used in commercial poultry houses to hasten maturity of young birds in order to get them into production as soon as possible, and also to keep them laying longer.

At the time of maturity, food intake increases, and the pullet, as she is now called, has a craving for calcium. This is usually supplied in the form of oyster shell grit. Calcium is needed to make the eggshells; and fowls also need grit in the gizzard to help grind up the food in the process of digestion. Remember that hens have no teeth! Feather growth slows down at this stage.

Eggs are usually laid in daylight, and at a later time each day until they get to the hours of darkness, when the hen misses a day and then starts another sequence in the early morning. As the time goes by, egg-production becomes slower, and the eggs develop watery whites and thinner shells. Ovulation ceases: no more eggs, no more sex hormones for a while. At this point the feathers start to grow again, stimulated by withdrawal of the sex hormones. Old feathers fall out, new ones grow: the hen is moulting and looks a sorry sight. If you didn't know what was happening, you would think your hen was sick.

Commercial producers need full production to make a profit, so they do not wait for the hens to go through the moult and start laying again. When egg-production drops to below half the normal rate, they either sell off the birds or force them into a moult by cutting down daylight length and depriving them of calcium. They used to withhold food and water, but this is stressful and is not recommended in welfare codes of practice.

After a rest and a new set of fine feathers, the hen becomes her old perky self again. She will now embark on another season of egg-laying. Backyard hens will lay for several seasons, and some backyarders buy hens that have done one season with a commercial producer. Given their freedom, they will soon learn to scratch for food, to flap their wings and to perch at night. It is rewarding to see them behaving naturally.

Backyard hens can't compete with the egg-laying performance of intensive poultry systems. Since seventy per cent of the cost of an egg is in the hen's food, they can't produce eggs as cheaply either, unless a source of cheap food can be found. But you can expect your layers to give you at least 200 eggs in a season. So, to work out the economics, add up the value of 200 fresh eggs and add up the cost of the feed needed to maintain the hen.

Diseases of laying hens

Free range hens do not suffer from disease very often because they are less stressed than their counterparts in battery cages. There are many health problems that can affect a flock, but many of them can be avoided with good management. This includes a balanced diet with access to clean ground. As with other livestock, observation is an important part of good management. So is hygiene, especially when the birds are restricted to a small area.

Professional poultry-keepers keep food and water vessels clean. Periodically they empty the house and, between batches of stock, the whole place is cleaned and sterilised. This should be done for home flocks about once a year.

Unless they have a very wide range, there should be at least two runs, so that one can be rested while the other is used, and this will help to control parasites. After use, the run can be raked over to spread the droppings mixed with lime to hasten their decomposition. If you buy blood-tested, healthy stock and you manage them well, there should be few problems.

Coccidiosis is one of the main poultry diseases. It is caused by a protozoa, and commercial producers medicate the drinking water of young birds to protect against it. In this case, good management means that the litter is kept clean and dry, because damp floor conditions can lead to coccidiosis. Prepared chick crumbs contain a coccidiostat, which reduces the effect of the disease, so that the young chick can build up its own resistance. Watery droppings, with blood in them, a miserable bird and sudden death in chicks are all symptoms of coccidiosis.

Birds that stop laying and lose weight may be suffering from a viral disease, leucosis, which causes internal tumours to grow, or Marek's disease, which can affect younger birds and sometimes causes paralysis of a wing or leg. For these diseases, there is no cure, and it is best to kill an ailing bird and to burn it, if possible, to lessen the spread of disease.

Vaccines are available to control fowl pox, and two more 'modern' diseases found in intensive units, ILT and IB, which are respiratory problems and not likely to occur in the well-managed backyard flock.

External parasites

In fact, parasites are the most common problem in small flocks. Scaly leg mite causes crusty deposits on the legs and can cause lameness. Scaly legs should be scrubbed with soap and water to soften the scales, and then painted with an oily substance, such as benzyl benzoate, which is also used to control mites in other livestock.

Red mites, found on the body of the bird, are blood suckers and a severe attack can cause anaemia. Sparrows can spread these mites and should be excluded from the poultry house if possible. Lice can build up in winter and louse powder will control them, as will dust bathing if the birds have access to some dry earth.

Internal parasites

Worms can affect fowls on free range, but the best control is proper feeding and a good rotation of runs, so that the parasite burden does not build up. If worms become a problems, medication can be added to the drinking water.

Eggs

Egg-production starts when the pullet is 17–24 weeks old, and a good producer lays an egg each day for about nine days, and then takes a couple of days off. The bird will lay for about six to eight months, and then stop and go into moult. After a rest of two months there will be a second season of egg-laying, lasting about ten months.

The first season is the best, which is why commercial producers usually keep birds for one season only, but hens can go on for several years, becoming gradually less productive.

You should get over 200 eggs a year from each hen, more if the birds are hybrids. Eggs keep quite well in the fridge or in a cool larder for weeks, except cracked eggs, which should be used first, or dirty eggs, which soon go bad. Once air can get into the egg, bacteria enter and the egg can deteriorate quite quickly.

It pays to store the eggs in some sort of order on trays, so that you know which ones are the oldest. Eggs should be stored point downwards on a tray, and they should be kept away from any strong-smelling substance, because the shell is porous and can absorb odour.

We used to store surplus eggs in 'waterglass', which is a solution of sodium silicate, and you may still be able to buy the powder at country general stores. In the solution the pores of the egg shell are sealed. Eggs will store in this way for several months, and will be quite suitable for cooking.

It is possible to store eggs in the freezer. Crack eggs into a bowl, beat up lightly and add a little sugar or salt to prevent thickening, depending on whether you will use them for savoury or sweet dishes. Or you can break each egg into an ice cube tray and when they are frozen, store the cubed eggs in a bag, which makes it easier to know how many you are using when following a recipe.

Faults in eggs

Soft shells This is normal at the start of the laying season, but at other times check that the feed is properly balanced.

Thin shells Shells are made of calcium, so provide oyster shell grit in the food for good shells.

Cracked eggs If the shells are thin, they will crack easily, but normal eggs will crack if there is not enough soft material, such as hay, straw or sawdust in the nest, or if the hens are disturbed.

Runny whites This is a sign of staleness, and occurs in old eggs, eggs stored in a warm place, or sometimes in very fresh eggs.

Pale yolks Not enough green food.

Blood spots in yolks Caused by small haemorrhages, sometimes due to a genetic fault, but it can also be the result of a fright while laying.

No eggs The hens may have come to the end of their laying period and be going into moult. Decreasing daylight length in the autumn usually decreases the number of eggs, which is why commercial producers use artificial light to increase 'daylight' length. Stress can also affect egg-production: undue disturbance, lack of food or water, extremes of heat or cold all may result in stress.

The size of the air cell in the egg tells you how old it is. When laid, the egg has no air cell, but air is gradually drawn in through the pores and the air sac gets bigger as the egg ages.

Horses

Horses can provide you with a lot of enjoyment, and a sensible, middle-aged horse can make an ideal children's pet. A horse is quite useful for keeping down grass in odd corners, and unless it is worked very hard, it will live mainly on grass, with a supplement of hay when grass is scarce.

Don't buy a lot of gear to start with. Wear a riding helmet of the approved SAA type. Riding boots are much safer than shoes. You don't really need jodhpurs, although you need to ride in long trousers. The current cost of buying horse and gear can be checked in the *Weekly Times* or in horse magazines.

Choosing a horse

All the proper questions should be asked before you start to look for your ideal horse. Ask yourself, 'Why do I want the horse?' There are so many ways to go, starting with a gentle hack round the local neighbourhood, which is my preferred form of riding. Your children may want to join a pony club, which could be an excellent way of getting to know people in a new area. The family may be sport-minded and want to play games such as polocrosse, or compete in eventing, dressage or show jumping.

It is wise to take riding lessons before you buy the horse, not afterwards (and your children as well, if they are to ride). Safe riding-school horses are the best kind to learn on, even if they are not very exciting to ride. After lessons it may be clearer what form of equestrianship you favour, and what sort of horse you want; or even whether you really want a horse after all. It is a great expense and a big responsibility to own a horse. It may be cheaper and easier to hire one occasionally.

Choose an establishment where learning is active, and not just following nose to tail round a trail with a group of beginners. Ask about the school so as to be sure that the instructors know what they are doing.

Don't buy a young horse as the family's first horse, because you will not be qualified to teach it manners. Buy from a reputable breeder, if possible, not from a saleyard. Look at advertisements in papers and farm and horse magazines, and shop around before you finally buy. You can learn a lot by visiting several establishments in search of your horse, so don't buy in a hurry.

It is, of course, essential to take an experienced horse person with you when you go to buy a horse. Take her advice. Even experienced horse breeders get caught out and buy the wrong horse sometimes. The price of getting it right is great vigilance. If it's at all possible, see the horse several times before making up your mind.

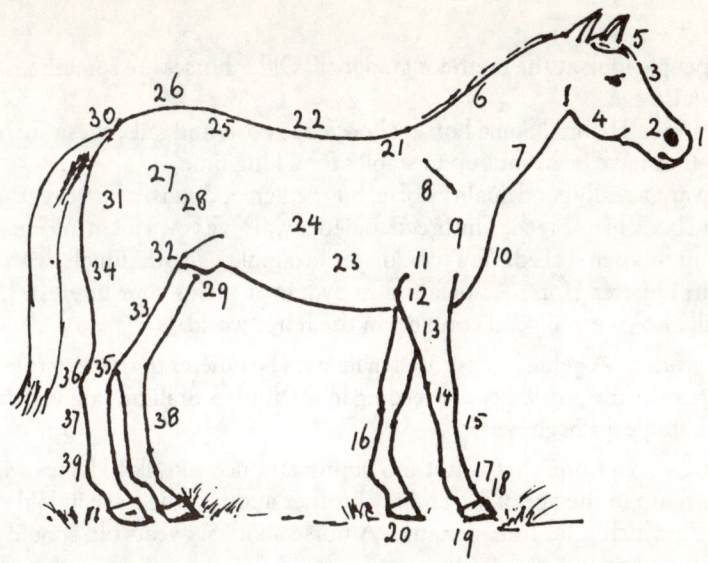

Points of a horse

1. muzzle
2. nostril
3. forehead
4. jaw
5. poll
6. crest
7. throttle or windpipe
8. shoulder blade
9. point of shoulder
10. breast
11. true arm
12. elbow
13. forearm
14. knee
15. cannon bone
16. tendons
17. fetlock/pastern joint
18. coronet
19. hoof
20. heel
21. withers
22. back
23. ribs
24. girth
25. loins
26. croup
27. hip
28. flank
29. sheath
30. root or dock of tail
31. hip joint round
32. stifle joint
33. lower thigh or gaskin
34. quarters
35. hock
36. point of hock
37. curb place
38. cannon bone
39. tendons

Checklist for horse buying:

- Will it be caught easily? Some horses waste a lot of your time avoiding capture. Beware of the horse that puts its ears back on its head and turns its rear to you when you go to catch it. We once had to sell a beautiful mare because she was hard to catch.
- Is it healthy, alert, with good feet and sound wind? On a trial ride, the experienced rider should put it at least to a brisk trot to see how it breathes when exercising. If it is lame, don't buy it. Watch out for a poor coat, split feet and a runny nose. If you decide to buy, get it checked by a vet first. This will cost money, but may save a great deal in the long run.
- Is it quiet? A horse should be quite happy for you to run your hands over it and pick up its feet for inspection. Ask the owner to handle/ride it first! It could have been doped to make it quiet for inspection. You could ask to take it home for a trial period of a week or so. The seller might not agree, of course. Or you could go back unexpectedly to see it again.
- Is it used to traffic, if that's one of your requirements?
- Is it the age they say it is? It is not always easy to tell with an adult horse, but

horse people look at the teeth for guidance. Older horses are sometimes sold with a false age.
- Has it any bad habits? Some horses chew fence posts and rails; these are often horses that have been shut up in stables for a long time.

If you want a really good quality horse, buy pedigree; that is, a horse registered in the stud book of its breed. The breed societies will give you a lot of information about your chosen breed. You could, for example, get in touch with the Australian Quarter Horse Association in Sydney if this is your interest. Local shows will also give you good contacts in the horse world.

The right gender   A gelding (castrated male) will be quieter to ride as a rule, but a mare presents the possibility of breeding in the future. Stallions are very frisky and not suitable for beginners.

The right age   Get one that is not too young and not too old! Horses will be quite active up to the age of 20, and my brother and his wife have an old mare of about 25 which is far from geriatric. A horse about six years old should be a well-mannered ride.

The right size   My favourite size for a horse is about 15 hands. (The height of the horse is measured from the ground to the top of the shoulder, called the withers, in spans of four inches. So 15 hands means the horse stands 60 inches at the shoulder.) A smaller pony might be better for children, and if you are very heavy, you may need a taller horse with a sturdy frame.

The right colour   There is an old saying that a good horse is never a bad colour!

Choosing the right breed

Take a look at the breeds and the jobs for which they have been evolved over the centuries. Technically, a pony is an animal under 14 hands, but ponies are rather different. Ponies are small, but often sturdy, and were used as pack horses and farm horses on small hill farms. There was a small dark Celtic pony in Britain before the Romans arrived, and the Exmoor is thought to be descended from this animal. Ponies are still used in hill areas of England for going round sheep flocks, since a shepherd can see more from a horse's back than on foot. Welsh ponies seem to be very popular round Melbourne, and they can be very flashy show animals.

Pony breeds are usually hardy and their temperaments are different from those of horses. They are more mischievous, and they can be hard to handle. The Shetland makes a wonderful children's pony, but in spite of its small size it can be very independent.

The Quarter Horse has been very popular in Australia for working with cattle. It was bred in Virginia to race in lanes near plantations. The races were short. The horses needed great acceleration to run a race over a quarter of a mile (hence the name). The horse was bred from Spanish stock, crossed with the English Thoroughbred. It was soon taken up by cowboys because it can accelerate and manoeuvre so well and it has the intelligence to know how to handle cattle.

The Quarter Horse has powerful hindquarters and is well muscled. The colour

Arabian horses are smallish, beautiful and intelligent. They were treated as part of the family when they lived with Arabs in the desert. In a famous painting by Landseer, the horse is lying on the carpet in the owner's tent.

can be any solid colour, with no patches on the body, but white feet and a white blaze (face marking) are common. It is medium-sized (from 15.2 to 16.1 hands) and makes a good riding horse, with a reputation for docility.

Housing

Do you really need a stable? Horses don't think so. A horse prefers to be outside. It will need shelter from the sun, from cold and hot winds, and maybe a rug in winter. If you rug it up, be sure to take off the rug regularly to check on its condition – that is, the amount of fat cover.

Fencing for horses is a subject for heated debate, but barbed-wire fences should be avoided if possible. Horses are relatively thin-skinned and are easily damaged by wire. They can be controlled by a low-voltage electric fence. Specialist horse-breeders use those fancy post-and-rail fences not just for show, but because they are the best for this animal.

If your horses, or indeed any of your animals, are really hard to keep in – if they break down fences deliberately and are desperate to get out – try to find out why they don't want to stay with you. They may be hungry or thirsty.

Breeding

If you plan to breed horses, learn their ways before accepting the responsibility of mating and foaling. Read some specialist books and talk to the experts. Like cattle, horses come into oestrus about every 21 days, but this happens only in spring and summer, September to January. So you can't breed from horses all year round. The period of oestrus during which mating can take place lasts five or six days. Mares are often mated straight after foaling, about nine days after the birth. The period of gestation is about eleven months or 335 days.

Horses need checking every day, like other stock, to see that they have plenty of feed and clean water. Don't keep the horse so far away from where you live that you see it only at weekends. This goes for other animals too.

Apparently, the number of farms with absentee owners and unsupervised animals is increasing. Some hobby farmers visit their farms only at the weekend. Don't leave the stock to their own devices. If you can't live there, arrange for an experienced neighbour to check on the animals every day. Arrange payment (help, produce or money), so that you can be sure the animals will be looked after.

With animals, things can go wrong quickly. Horses live on grass, but they are not ruminants, so they have lengthy guts. Anxiety or stress can cause colic, and so can mouldy hay, foreign objects in the feed or internal parasites. Any digestive upset can cause colic. Although it is simply stomachache, a severe bowel pain, colic can kill a horse if it is not treated immediately. With colic you have to keep the horse moving and, several times, I have had to walk a suffering horse round and round the yard while we waited for the vet.

External parasites such as the mange mite can cause itching and loss of hair. For beginners, it is a good idea to have your horse regularly inspected by a vet as part of a preventive health programme.

Feeding

The horse's condition is a good guide to the adequacy of the feed, so watch it carefully. Horses that go out a great deal and are ridden hard will need some grain feed, usually oats. But it is dangerous to feed a horse too well, so don't overdo it if the work is only light. Up to about 8 kg of oats, usually mixed with lucerne chaff, can be given per day in two or three small feeds, with a little calcium added. A horse weighing about 500 kg will need about 14 kg of lucerne hay to maintain itself without working.

Too many oats may make the horse excitable and hard to handle, and too much lush grass can cause it to 'founder', in which condition it is painfully lame with a swelling under the hoof. Horses should never be given mouldy hay. Any changes to the feed should be made gradually.

Horse care

Flies can cause intense annoyance to horses, leading to sores when they rub the irritation. Some people treat the animal with fly repellent, or you can use a cotton rug. Mosquitoes can cause an allergic reaction. The best way is to eliminate fly-breeding sites, if you can, which will be a good thing for everybody else on the holding.

Regular grooming is good for a horse, but unless you aim to compete in dressage events, the animal does not need to be constantly washed. Brushing is something the horse should enjoy, and it will help to keep the coat in good condition. The close attention that brushing entails will ensure you spot any skin problems early.

Well-fitting saddles and bridles are, of course, essential for happy horses. Be aware of any problems your horse has with the gear: check the back for saddle marks after riding, and be sure the mouth is not sore. (Sometimes the teeth may need filing – ask the vet.) Buy the saddle and bridle *after* you buy the horse, to be sure of getting them exactly the right size.

Internal parasites

Mixed grazing, say with horses and cattle, so the horse can fit in with your other stock, is best. But watch carefully to make sure that they get along well. Many horses like to chase cattle, especially if they have been bred for the job.

Horses and cattle will interrupt the life cycle of each other's worms. Even so, horses should be rotated into different paddocks to give the parasites time to die down. Worms in animals are much more of a problem now than they used to be: most horses are regularly drenched for worms, and you should ask your vet for advice. The long-term solution to worms is to work towards healthy horses and healthy soil, so that the animals build up a natural resistance to parasites.

Care of feet

Footcare is obviously most important. When shoes are worn, they will need replacing regularly, and unshod horses will need regular foot-trimming by an expert. Some horses have hard feet, which need little trimming, but others can split if they are not trimmed often. Make sure that the farrier likes horses, is patient with them, comes well recommended, and is registered with the Master Farriers Association. Wise horse-owners make footcare enjoyable for both the horse and the farrier.

Some horses hate the farrier, but there are things you can do to stop them associating footcare with fear. We always get the farrier to come to the horse rather than the other way round, because horses are better in familiar places. The action takes place in the yard, not in the stable, because the stable is the horse's home and refuge. We try to exercise the horse before the farrier arrives so that it is not too impatient to get away. If a horse has a companion (our old Whisky is lost without its mother!), have the companion nearby to keep the horse calm.

We usually stand at the horse's head and talk to it during shoeing. (There is a reward afterwards too.) Some horses need oil on their hooves to stop them from drying, which makes them more susceptible to cracking.

Pigs

Why keep pigs?

There are good arguments for keeping a few pigs as part of an integrated backyard system, although it must be admitted that they breed quickly and that too many pigs can wreck the system. Too many pigs are much worse than no pigs at all.

What will two young pigs do for you? Several things:
- Part of the garden can be dug and fertilised by pig power. They will uproot and eat problem plants like bracken and blackberries and are very good for weed control on neglected land. Pigs are the best help you can have in starting a new garden.
- They will convert waste food to good meat.
- Surpluses from a dairy enterprise or from the garden can be used as pig food.
- They are interesting animals to keep.
- Pig manure is a valuable fertilser or compost additive.

To choose the right type of housing and management, go back to the habits of the wild pig and the early domesticated ancestors. Pigs have no protective fur, and in the wild they creep into dense undergrowth to sleep and to get away from extremes of weather. Their young are born in nests made in safe undergrowth.

The early domesticated pig was a forager, keeping down the rubbish round human settlements by eating all edible scraps. Herds of pigs were taken into the forest in the autumn to feed on the seeds of beech and oak trees. At other times of the year, they fed on weeds on the fallow (the land temporarily 'resting' on rotation, out of cultivation) and the grain left behind after the harvest. Pigs enjoy an outdoor life with plenty of variety and exercise.

The pig's digestive system is similar to that of people. Like us, they need roughage, but can't handle too much of it. They will eat grass, but do not digest the cellulose. They need variety, and this is something that the backyarder can provide. Commercial pigs get a ration of barley meal (or maize in the USA), with added protein, vitamins, minerals and sometimes growth promoters. The bland cereal formula must be very boring to the pigs, as is the whole indoor pig-keeping system.

Choosing a breed

Start by talking to pig-keepers and looking at breeds, but don't bring home your pigs until you have done your research and can provide the right kind of home for them. This is particularly important with an animal that is powerful, intelligent and destructive. Pigs are quite good-humoured, as a rule, but they can push with their noses and put a lot of weight and ingenuity into escape bids.

The ancestors of our modern pigs were ferocious European wild pigs living in the forest; they were probably domesticated by Neolithic people. This rough-looking animal was crossed with Asian swine, and so our pigs owe something to the historic pigs of China. Chinese pigs were smoother, rounder and more like the modern pig to look at, although probably rather smaller. Pig meat was important in the Chinese diet, and the pig manure helped the small farmers to keep their land fertile and in good heart.

As with poultry, the modern mass-produced pigs are hybrids and tailored to an artificial environment. Try to buy your weaners from a herd that lives out of doors; and go for one of the old breeds if possible. It may be wise to choose a breed that is already kept in your area, especially if you want to try pig-breeding and will need the services of a boar.

Rare breeds of pig are gaining in popularity now. For example, the Vietnamese pot-bellied, a small pig, is well thought of by New Age farmers in England. There was a lovely old spotted breed from Gloucester, and a ginger pig called the Tamworth. This was brought to Australia because of its resistance to sunburn.

Duroc, Large White, Landrace and Hampshire pigs were advertised in the *Weekly Times* when I checked recently. A brief description of some of the pig breeds follows.

Large Whites
A good bacon pig, often used for crossing to produce hybrids. Originally bred for backyard farming by Yorkshire weavers, it is called the Yorkshire in the USA. They are inquisitive pigs with prick ears. The lop-eared breeds tend to be more docile, because they can't see as much! The Large White and its cousin, the rarer Middle White, is a good mother and a good forager, but can be clumsy with the baby piglets and sometimes aggressive in their defence. I have been chased out of a pig-pen several times by a Large White mother in an unsociable mood.

Duroc
Popular in Australia, this breed originated in New Jersey in North America. Tough Durocs are suited to an outdoor life: their dark-red colour makes them less susceptible to sunburn than the white breeds.

Landrace
There are Danish, Swedish and Finnish variations on this Scandinavian breed. They are white, long and lean, with flop ears, and they are very docile. Two older breeds, the Welsh and the British Lop, are similar but not so long.

Hampshire
This is a very old breed, the original of which was the Wessex Saddleback from the south of England. The Americans bred it up in Kentucky in the late nineteenth century and renamed it the Hampshire. It is my favourite breed of pig, black with a belt of white round the middle, not so big as the Large White, and a good outdoor animal.

Pig-keeping systems

Pigs are curious and intelligent animals and need to have something to occupy their days, unlike ruminants, which spend most of their time grazing and then sitting down to chew the cud.

The secret lies in giving the pigs what they need, while still considering the needs of the land, the other stock and people. On most smallholdings, this is done by a combination of indoor/outdoor pig-keeping. Sometimes they are on the land, sometimes off it, but always in harmony with the system. I have kept pigs in a variety of systems.

An A-frame hut, built sturdily of wood and with a wooden floor, can be made from scraps of timber; or the hut can be made of corrugated metal or even straw bales, netted to prevent their destruction by the tenants.

A movable hut gives you flexibility, because the pigs can be put onto whatever piece of land needs to be dug and fertilised.

Pigs will normally keep the hut clean, making a toilet area outside. They will need straw, hay or other bedding, because they like to burrow into it to sleep. The fencing will need to be good. Pig wire, buried to prevent escapes from under it and strained tightly to wooden posts, will confine the pigs, and little pigs should not be too much of a problem.

The pigs should not stay on one plot of land too long. This is because there may be a build-up of parasitic worms which could affect the health of pigs. The

Outdoor pig This pig is not deprived of liberty and can run about and behave naturally. It also has a shelter to protect it from heat and cold. Pigs need protection because they have no fur to keep them warm, and also for security. Wild pigs sleep in deep undergrowth at night and our pigs feel more secure with a darkish shelter in which to retire.

other reason is land care. In extremely wet weather, the land will become a mud bath; in dry weather, a dust bowl. Either can cause loss of soil structure. When all the vegetation has gone, it is usually time to move on.

The way we solved the problem of land care was to bring the pigs indoors for some of the time, or to position the movable hut next to a fenced concrete slab at certain times of the year. From the concrete yard, the pigs could survey the world and take fresh air without ruining the paddocks.

Sometimes we kept sows in a 'cottage sty', which was a low hut with an open yard attached. This was a traditional building in English cottage gardens, and our family used it to good effect when we farmed in the mountains of Wales, where rainfall is high.

If you have very little space for outdoor pigs, they could be kept indoors in a converted garage with plenty of straw and green food, and diversions such as chains to play with. Then perhaps you could let them out in a grass paddock for exercise while you are there to watch them. They can soon be trained to come back inside at the rattle of a bucket! The bucket is the best form of control for most domesticated animals. Get them used to being given food from a bucket, and they will follow you anywhere.

When we had paddock space but wanted to preserve the grass from being dug up by the pigs, we sometimes put a ring in each pig's nose. In this way the pigs could enjoy the outdoors and eat the grass without destroying it. This may seem cruel, but to my mind it is kinder than keeping pigs indoors.

In the more natural systems of livestock-keeping, the climate is probably the main consideration. If you buy weaners to fatten, they will be with you for six months at the most, so you can choose the best time of year for the pig project, perhaps the time with the most surplus food from the garden or the most crops for the pigs.

Homesteaders in America usually prefer to keep their pigs in the summer months, especially in states where the snow is deep in winter. Many homesteaders grow corn especially for the pigs, since corn (maize) is the easiest cereal to grow in small quantities and it can be fed on the cob. This is good food for poultry as well.

Australian backyarders might find summer pig-keeping the best in temperate zones, but winter will be easier in the northern states. Of course, commercial systems have their own controlled environment and run all the year round. If you breed your own pigs, the sow will be with you all year.

Pig-keeping on small areas

There was a time when most country folk kept a pig or two, fattened in a sty at the bottom of the garden on kitchen waste and weeds. The pig was the main source of meat for the family, an important part of self-sufficiency in the days when people grew their own food from necessity. Although it was to end up on the table, the pig was a family pet and the centre of conversation when neighbours came to visit.

Small-scale pig-keeping is out of fashion now. Now the pig industry is in the hands of specialists, pigs are reared in their thousands, and backyard producers are few and far between, although you can still find a few if you look for them. Pig-keeping is probably not suited to suburban gardens in these tidy times, but there is still a place for a couple of pigs on a small farm or on a half-acre plot on the city fringes or in country towns. The best way is to start with small pigs and watch them grow. They are interesting animals, and will make a good contribution to your backyard system.

Buy two weaner pigs (they need company) of the same sex, unless the male is castrated. If you are thinking of breeding pigs, you could buy two young females or gilts and keep one for breeding. At six to eight weeks old they should weigh about 20 kg each and will have been weaned onto solid food. You can then, without too much trouble, fatten the pigs for pork or bacon. Pork pigs are ready at about five or six months, when the joints are the right size, and bacon pigs are rather larger and older; say about eight months on a mixed diet, although the specialists grow them faster than this.

Feeding

At least three-quarters of the cost of conventional pig-keeping is in the food they eat, so this is the main area in which savings should be possible. As I said earlier, commercial pig rations are based on grain, with various additives. As a beginning, you might like to try feeding them partly on commercial rations, increasing the forage or cheaper foods as the pigs grow in size.

At weaning, when they are bought in, the little pigs will eat about a kilo of cereal food each per day. This can come as dry meal, pellets or cobs, or meal mixed with water to form a gruel. Large cobs are good for outdoor systems, because they are too big to be carried away by birds. We feed pigs twice a day, morning and evening, but you can cut this down to once a day for older pigs, who will forage for some of their own feed.

A good practical method of gauging feed is to give them as much food as they will eat up completely in twenty minutes. The amount is increased gradually as the pigs grow, until they are eating about two kilos a day of cereal feed or equivalent. Don't overfeed them, as this will ruin their digestion!

When vegetables, weeds and so on are used as a substitute ration, the bulk of feed will need to be larger, as the protein and energy value of the feed is less than that of a grain ration. Little pigs cannot deal with bulk very well, but their capacity increases as they get older. So the ration can be bulkier for older pigs.

You may be able to get small potatoes from the growers. They should be cooked before feeding to the pigs to get the best value from the starch: 4 kg of spuds are equal to 1 kg of barley in feeding value.

You can judge by the condition of the pigs whether they are getting enough: they should be rounded and firm, shiny with health.

Pigs can fit into agro-forestry systems with care. Our family kept pigs in a mature apple orchard, and they did little damage to the trees, eating grass and windfall apples and pears in season: occasionally they got drunk on fermenting apples. Forage trees can be protected by fencing, and the pigs can pick up what falls or leans over the fence.

Many different crops can be grown for pigs, again depending on the climate. In a high rainfall area, comfrey, with its fast growth of high-protein leaves, will be a good pig crop. Comfrey is a neglected source of protein, and it has many uses. I use it as a garden fertiliser, a healing ointment and animal food. Stock sometimes take a while to get used to it, but once they take to it they love it.

Roots such as turnips, carrots or fodder beet are traditional pig fodder, and while the pigs would enjoy digging out roots for themselves, it is probably less wasteful to cart them to the pigs.

Breeding

Breeding from a newly acquired pig

The easiest way to start breeding your own pigs is to buy a pregnant female. Give her at least a few weeks to settle down in her new home before farrowing. The sow will need a clean place for farrowing, preferably indoors, so that you can keep an eye on her. She will need clean straw, because the sow has a strong instinct to make a nest before the arrival of the piglets. In fact, when she starts to run about with straw in her mouth or scrape it into a heap in a dark corner, the litter is on the way.

Milk will appear at the teats and the sow will be restless. Then she will lie down and give birth. Keep an eye on what happens next, but don't interfere unless she needs help. As the piglets are born, the cord breaks and they struggle round, by trial and error, to find a teat to suck. Sometimes a sow is so restless during farrowing that it is best to take the piglets away and keep them warm under a lamp until she's finished, then give them back to her. Other sows are better with the piglets around.

Nervous sows can be aggressive at first with the little pink strangers, and the traditional trick was to give them beer as a tranquilliser. Pigs love beer, and it does calm them down. A good sow will manage the piglets very well, especially if she can go outside. The babies will soon be collecting their own minerals from the soil: indoor piglets need injections of iron, but outdoor pigs find their own. Dose with iron by opening the mouth with your finger from the side and smearing paste on the tongue.

After the piglets are weaned, the sow will come into oestrus again and can be taken to a boar. The gestation period is nearly four months for pigs, so the first litter will be about ready for pork by the time the second litter is due.

Whenever you handle baby pigs, keep an eye on the sow. She may be protective and try to bite you. It's a good idea to give the sow some special titbit when you handle the litter.

116   Animals for Gardens and Backyards

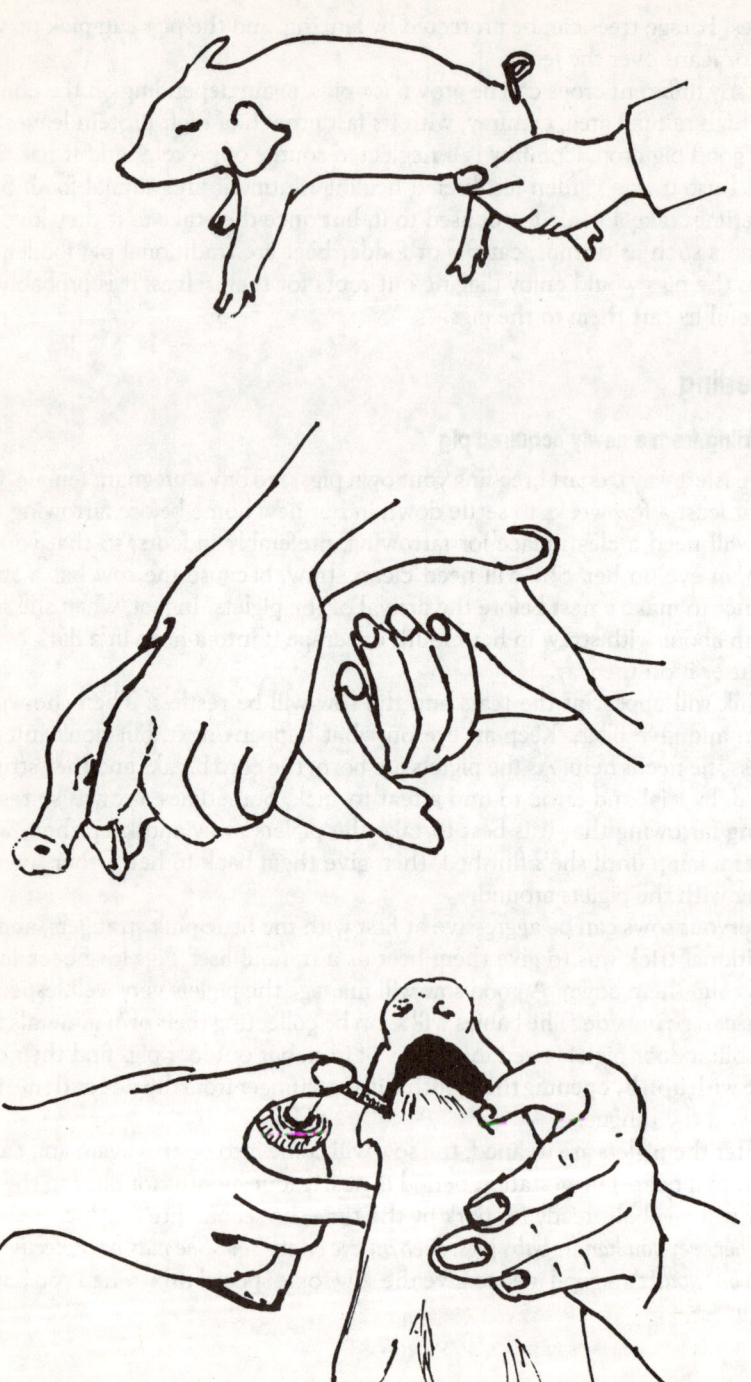

1. Catch a baby pig by the leg and lift it gently, supporting the body with your other hand. If you grab it round the ribs it will squeal loudly and the sow may attack you.
2. Indoor pigs need extra iron, and easier than an injection is the paste on the tongue. Insert your finger gently in the side of the mouth to open it.
3. With the mouth held open, squirt the paste onto the tongue with the other hand. Hold the piglet in the same position for a moment so that the dose is swallowed, then replace it gently with the litter.

When treating baby pigs it is difficult to tell which ones have been treated and which haven't. We used to put them in a barrel after the dose and return them all back to the sow at once. Or you can give them a dab with a marker. There is a special crayon for livestock use.

Breeding from your own pigs

A gilt or young female is ready for breeding at six to eight months of age, when she gets to the size of a bacon pig. If you decide to keep one for breeding, choose a nice-looking animal with at least twelve teats and a quiet disposition. The number of teats varies, but she should be equipped to feed a large litter, since piglets tend to claim their own teat and stick to it.

By this time you will probably know pig-keepers in your locality and will be able to take her to a neighbouring boar. Pigs come into oestrus for about three days at a time at three-week intervals and, when ready for service, your gilt will stand perfectly still and be almost impossible to move.

The easiest arrangement will be to have your pig run with the boar for a while, or stay at the farm for a few days so that she can be served twice during the heat period. The owner of the boar will supervise the event for you. Four months later, the piglets will be born. The new mother will know you and trust you, so there should be few problems with her behaviour, but pigs are unpredictable when they have young, so watch out for some hostility at first.

A sow loves having her belly scratched; it soothes her. If you tame her to lie down when you scratch her, you will be able to soothe her when she is giving birth and maybe make it easier for her to accept the babies.

Give a pig a few feeds of bran before farrowing. As with an imported female, the pig will need somewhere clean and safe to give birth, with the chance to make a nest. Outdoor pigs can be brought into a clean pen with plenty of clean straw, or they can stay in the pig shelter, providing it is warm enough for the babies and that you can get in there too.

Sows know what to expect, but a young gilt can take fright when she sees her first piglet. Sometimes she will let the babies suckle, but can't bear to see them. Piglets can be lost by a hysterical sow, which might savage the babies or trample on them. In fact, the quietest sow can squash piglets by mistake when she flops down to feed them, because they are tiny in relation to her bulk.

The birth process is not normally too hard for pigs because the little ones are so small, but sometimes two piglets try to come together, or there is one wrongly presented. So it is as well to keep an eye on the process.

With a nervous sow or one that was restless, we used to stay with her at farrowing, taking the babies away as they arrived and keeping them warm under an infra-red lamp until she had finished. You can expect nine or ten . . . or twelve or thirteen!

Piglets are born in a membrane and they sometimes suffocate before struggling free, so you can save lives by being there to clean their faces gently and encourage them to breathe.

If a sow strains for a long time with no result, there could be a blockage, and experienced handlers disinfect hand and arm and then investigate. But this is only an emergency procedure, as any interference can cause infection. In spite of all these alarming suggestions, most pigs will quietly farrow by themselves, feed their piglets and get on with the job of mothering. In a few days you can let them out. But make sure the fence is piglet proof!

Weaning

Weaning used to be at about two months of age, but commercial herds now wean earlier. The best way is to let them decide: wait until the piglets are eating solid food and drinking water, and the sow is maybe getting thin and needs a rest.

To help the sow out, especially when there is a large litter, you can 'creep feed' the young ones. Arrange small pots of clean water and sweet-tasting cereal (or commercial piglet pellets) in a place where the babies can get through, but the sow can't reach. This is called a 'creep' and it will encourage the piglets to eat so that they are not completely dependent on milk.

Once the piglets are weaned, they will be able to eat more roughage, but introduce it gradually. They will perhaps weigh 20 kg when weaned, so kitchen scraps and cereals will be suitable for them. As they grow bigger, they will be able to handle more grass, weeds and vegetables.

Pig health

Overfeeding used to be the main cause of unhealthy pigs in the days when people wanted fat bacon. Since then, pigs have got leaner, and the main problems now are overcrowding and artificial living conditions, boredom and depression. The backyard pig should not have too many health problems, if given adequate food, water and shelter.

Scours is the main problem encountered with newly weaned pigs. *E. coli* bacteria can cause scour where the pigs are stressed, for example by weaning and moving to another location, and this is probably the most vulnerable time for pigs.

A clean and warm sleeping place with plenty of straw will reduce the stress on pigs of any age. If piglets are scoured, fluid replacement is important, or they may die from dehydration. Electrolyte solutions can be added to the drinking water, and the vet can supply probiotics (the opposite of antibiotics), which replace the harmful bugs in the intestine with beneficial ones. In the absence of probiotics, you can use live yoghurt culture to do the same thing. Too much food may produce a nutritional scour. Be careful not to overfeed, especially young pigs. But if scours persists in older pigs, call the vet.

Infectious diseases are less of a risk with outdoor pigs, who should develop a healthy immunity, but there is always a risk of disease. Change your clothes after visiting other people's pigs and discourage visiting pig-keepers from handling your animals. Diseases include erysipelas and rhinitis.

External parasites

These can cause trouble to pigs. If the pigs appear itchy and rub themselves on posts a great deal, examine them for mange or lice. There are some powerful chemical treatments for parasites, or you can try the organic ones. Good, clean housing and correct feeding should minimise the problem.

Internal parasites

Worms may be a problem where pigs are left on the same piece of land for too long. If the pigs grow slowly and cough a lot, they have probably got worms. Wash the pigs thoroughly, put them indoors, and clean them out every day as a control measure, with plenty of garlic in the food. The worms are picked up from manure, but if they are kept clean indoors, the pigs should avoid worm problems. There are several herbs in addition to garlic that make good wormers: tansy is one, but it will have to be added to the food, as pigs will not touch the plant if given any choice. To prevent this sort of problem, keep the pigs moving. If possible, put them on fresh land every month or so.

Heat-stroke

Pigs transported in hot weather and sows about to farrow are the most susceptible to heat-stroke, but it could happen to your backyard fatteners if they are out in the sun without shade. Make sure that they can get away from the sun when they need to, and if they do seem too hot, a hosing down will probably revive them.

Bringing home the bacon

Killing the pig used to be a ritual for the whole community, and one member of the group specialised in doing the job. These days, an on-farm butchering service will drive up with a mobile coolroom, and everything will be done neatly and professionally, and hygienically too. It is quite legal to kill stock on the premises, so long as you use it for home consumption and do not sell it.

There are several reasons why killing on the farm is preferable. Most important is the animal's welfare: the animal is not subject to the stress of a journey to the abattoir. It makes for a quick and humane end for the animal, and also for better quality meat. Another reason is that if you have the job done at home, you are certain that the meat you put in the freezer is that from your own animal. The on-farm butcher can of course be used for sheep, goats and cattle, as well as pigs.

People often have qualms about eating home-produced meat, but by the time the butcher has packaged it for the freezer it looks just like supermarket meat, which often solves the problem.

The weight and fatness at which you kill the pigs will depend on your preference, but most people prefer small and lean joints these days. Pork pigs, when ready, will weigh from about 40 to 70 kg live weight, which will give you 30–50 kg dressed weight. Take away the head and some of the larger bones, and you will end up with perhaps 25 kg of meat from the average pig, not too much for the family freezer.

Bacon pigs are bigger, perhaps from 70 to 110 kg live weight. Home-cured bacon is a useful aid to self-sufficiency, but it is hard to get the cure just right. To imitate modern bought bacon, you will need to cure the bacon in brine. The old-fashioned dry salting can work quite well unless the weather is warm and humid. The bacon can easily be tough and salty as a result. My brother and I did quite a lot of bacon-curing when we bred pigs, but not always did we get it right. Mind you, it never went bad; and we stored the bacon in a muslin bag, hanging from the ceiling in the old-fashioned way. We kept the side of bacon, or the ham, covered in salt for several weeks, turning it every day, before drying it and suspending it in the bag.

Brine salting or sweet pickling is probably easier for a modern household. You can keep everything tidily hidden in a plastic bin. A brine mixture can be made by mixing up 18 litres of water, 3 kg salt, 125 g saltpetre, and 750 g brown sugar. Boil the mixture for about twenty minutes, then cool. Soak the bacon in the brine for several weeks, or pour it over the meat every day. You might like to experiment with a small joint to begin with!

There are plenty of sausage recipes in the European tradition for you to try, using the scraps of pork after jointing is finished. The English tradition was to make a pork pie of the small bits, using hot water pastry: 'raising a pie', it was called. They also loved to make 'black pudding' from the blood. In the days of the cottager's pig, nothing at all was wasted: even the meat from the head was made into 'brawn' or 'head cheese'. The pig was shared by the whole village, and the owners were repaid when somebody else killed a pig.

Rabbits

The rabbit is a native of Mediterranean countries, and of course Australia suits it only too well. Rabbits need a warm, dry place in which to sleep and a nest in which to rear the young.

Domestic rabbits have been bred for meat, fur or wool (in the case of the Angora). In fact, most backyarders may think of rabbits as children's pets, not something to make into a casserole. In this case the Angora will certainly be the best breed and may introduce the family to a fascinating new hobby as well.

Regulations are important in a country where the wild rabbit is a menace. Two domestic rabbits and their litters are acceptable in the local authority area I investigated, and this would be enough to supply you with enough rabbit meat for a family, or some Angora wool for spinning. Check with your authority what the local laws permit. Rabbit-breeding is quick. Young does are ready to breed at five or six months, and each female will produce several litters a year, with five or six young a litter, so one doe will probably be all you need.

Commercial rabbit farming is not allowed in all states of Australia. In 1987 it was first allowed in Western Australia, with restrictions as to the type of enterprise. Licences were needed, and Angora breeding stock was imported from Germany and New Zealand.

Housing

The usual rabbit hutch does not give the rabbits enough chance to exercise. The problem is that they are burrowers. If you give them a stretch of earth, they will dig their way out. A compromise is to let them run loose in an old building such as a garage or a stable. Or you can make them a movable ark.

An ark suitable for rabbits would be an A-frame hut with a run attached. The floor would be covered with wire mesh. This give the rabbits a chance to eat grass whilst it stops them from digging. The ark can be moved over your lawn, and the rabbits will do the lawn mowing for you.

This type of ark is very useful for all kinds of small livestock. My brother has several arks in which he brings up young turkeys, letting them have more freedom when they are old enough to look after themselves.

A hazard with this type of arrangement might be the risk of disease from wild rabbits.

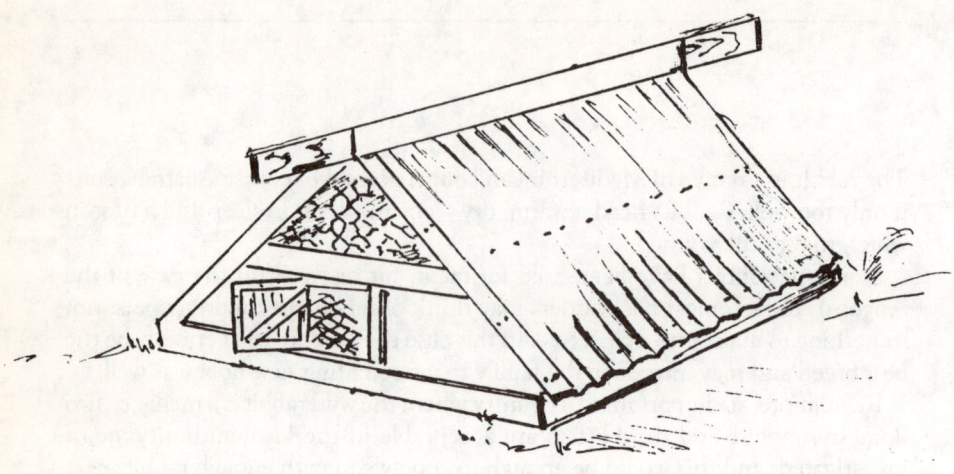

Movable hutch

Hay racks like this, made from wire in a sort of pouch, can be made for all kinds of stock. This one is rabbit sized and low enough for a baby rabbit to experiment with eating solid food.

Racks are important because hay left on the floor is soon treated as bedding and will get dirty. Goats especially will not eat anything that is not perfectly fresh, bless them. You need good racks for goats, and for sheep and cattle they are useful too.

If you make detachable rabbit racks they can be filled outside the pen.

Breeding

At one time we bred meat rabbits and we found that they were not quite as prolific as their reputation would suggest! Depending on the regulations, the backyard rabbitry might have one buck and up to five does. Most breeds of rabbit are presumed to be always in season, although not all breeders agree with this. Angora breeders think that the doe has an oestrus cycle of something like three weeks. To pick when to mate them, try a rabbit which is alert rather than resting. The doe is always taken to the buck's pen for mating, because the doe will make war if any other rabbit comes into her territory. We got quite a few scratches from handling excited rabbits at mating time, so take care. Supervise the pair while they are together; if you leave them, they may fight and injure each other.

After mating, you must wait a month for the litter: the gestation period is 28–31 days. Before the due date, everything must be ready. Rabbits make a nest for the litter, so you have to give the pregnant female the right surroundings and materials. A few days before the birth, we prepared a wooden nest box with fine chopped hay or clean shredded paper in it. The doe will probably arrange the nest with a bare floor, so the nest box needs a warm floor. If she is to give birth in the pen, it should be cleaned and fresh litter supplied. The doe plucks fur from her front to line the nest. She will probably eat less food for a few days before the event. Rabbits are best left alone when giving birth. They seldom have trouble, and they clean themselves up afterwards. But things can go wrong if the doe is nervous through having been moved or upset, and then she may drop the young anywhere and they can chill and die. Five or six babies is about the norm. You can perhaps distract her with a special feed while you have a look at them. They are blind, deaf, and naked and helpless little things at first. Angoras are not particularly good mothers as a breed, and may rear only half the babies to weaning. As the babies grow, the mother will need more food. At about fourteen days the babies' eyes will be open, and they will be growing fur. By about three weeks they are venturing out of the nest, and you will see them experimenting with solid food.

Rabbits, especially does with a litter, need a lot of water. Make sure they can get water when they start to eat, but not in a dish they can drown in. You can make a hygienic drinker by filling a bottle with water and upending it in a shallow tray: keep your finger over the neck until it is under the water in the tray. A board with a wire ring at the top will secure the bottle. This kind of drinker is also suitable for small numbers of poultry, and it is safer for young rabbits and chicks than a deep container.

Weaning depends on progress more than age. When the young ones are independent, they can be weaned without stress, say at about six weeks or so. Most rabbit litters are born in the spring, summer and autumn. Shorter daylight hours lowers reproductive activity in the rabbit.

Rabbit manure

One doe and her litters will produce about six cubic feet of manure a year, according to American research. I have not measured it, but I know how useful it is. We put it in the compost heap, but found that you can also use it directly in the garden without composting. Chemical analysis of rabbit manure (approximate percentages) is: nitrogen 2.5 %, phosphorus 1.5 %, and potassium 1 %.

Commercial rabbit-producers sometimes have a problem with odour, which can be quite strong in rabbit manure. Many of them solve the problem by combining the rabbitry with a worm farm, since rabbit manure is ideal food for worms. Worm-pits are built underneath the rabbit cages, which have wire-mesh floors to let the droppings fall through. Backyarders may not want to keep their rabbits in such small cages, but there is still the possibility of breaking down the odiferous manure quickly if you feed it to the worms.

Rabbits for meat and fibre

American homesteaders are quite keen on keeping rabbits for meat, because they see the animal as an efficient producer of protein. Rabbit meat is high in protein, 20.8 per cent protein compared with beef at 16.3 per cent. The fat content is 4 per cent, which is lower than most other meats, including chicken, and the mineral content is the highest. The meat is white and milky in flavour.

Feeding for rabbit meat production

Oregon University has a rabbit research centre. Their research in the 1980s suggested you should not rely solely on grain-based commercial rabbit pellets. They successfully replaced grain with a meal made from lucerne (alfalfa meal).

Backyarders can feed rabbits lucerne hay and high-protein plants like comfrey to give them the balanced diet rabbits need. Recycling of kitchen vegetable scraps, stale bread and porridge leftovers is much better for the planet than using grain to feed animals: grain is better fed to people direct. Under this system, the rabbits get some of the grain that people leave.

The problem with feeding all greens is that they are over ninety per cent water, so the rabbits need a great deal to get enough nutrients to grow quickly. Root crops, such as carrots, are also high in water content. Beware of grass clippings: if they ferment, they can make rabbits very ill.

With a meaty breed such as the New Zealand White (the most efficient producer of meat) or the Californian, which has black ears, nose and feet, production is quick with the right diet. In eight or nine weeks the litter will be 'fryers' weighing about 2 kg, over half of which will be meat.

Angora rabbits for fibre

This is a long-haired breed of rabbit with large, floppy ears. It is one of the oldest breeds of rabbit, probably kept for its wool by the Romans. It appeared on Roman coins in the time of Hadrian. Some people think that, like the Angora goat, it came from Angora in Turkey.

Angora rabbits are very far removed from their wild ancestors and would probably not survive in the wild. They would not be able to run fast enough to avoid danger, and would probably get their long hair tangled in the undergrowth.

Feeding angoras

A rabbit on commercial rations gets about 250 g of high protein pellets a day, a pregnant doe twice as much. Rabbits don't eat very much. Backyard rabbits will thrive on a diet of fresh green food with a little hay in a rack always available for them, and a bran mash made from kitchen scraps such as stale bread or other cereals.

Rabbits are vegetarians, of course, and will never eat meat. When you collect wild green food for rabbits, remember that they are Europeans. Give them introduced weeds such as dandelion, plantain, chickweed, fat hen, groundsel, and so on. Shepherd's purse and yarrow are tonic plants for rabbits. Try to vary the diet; do not give them too much of one thing, and remove stale food from the trough. Cut down if they do not eat all their food. Water is very necessary to all livestock and rabbits need a constant supply of clean fresh water. At pet shops you can get nipple drinkers which supply water without flooding the pen.

Health

A healthy rabbit is attractive, with wide open, bright eyes, a good appetite and an active disposition. The body is firm and well covered, the fur is even, and the faecal pellets are hard. The nose should be dry and the ears clean. Get to know the rabbits when they are well, and then you will notice any early warning signs of ill health, such as half-closed eyes or a tucked-up posture. Sneezing means respiratory problems, which are quite common in intensive systems. If a rabbit shakes its head a lot, it may have ear canker.

Buying in rabbits should be done carefully from disease-free stock. Rabbits on a natural system should not suffer from the snuffles and sneezes that beset the intensively kept commercial herds. Pneumonia is a hazard among housed rabbits. Digestive troubles can be scouring or the opposite: impaction. Plenty of green food and fibre in the ration should keep their digestions working properly; but a sudden change in diet can cause scour, and so can wet greenfood. Comfrey is a healing herb, which is very good feed for any animal with digestive problems. Too much green food can cause a condition rather like bloat in ruminants. Sometimes rabbits suffer when their teeth grow too long, and you should never breed from those with overgrown teeth. Cases of sore hocks are sometimes seen when the rabbits are kept on mesh, or are left on solid floors that are not cleaned out often enough. Heat-stroke can occur in very hot weather: the rabbit will breathe fast and lie on one side. Cold water should be available for them, and you could place the rabbit on its side for a while on crushed ice over which some hay has been spread. The main thing is to minimise heat-stroke by arranging plenty of shade and plenty of space for each animal.

Sheep

Could you get to like sheep? Sheep have been kept for thousands of years for family use and they are docile and pleasant to deal with. Small and easy to handle, they are kept on both smallholdings and sheep stations for meat, wool, and milk. They are a wonderful animal for small-scale farming, although this is not their usual role in Australia.

The typical Australian sheep is a wide-ranging Merino, not exactly like a backyard sheep. Merino sheep don't enjoy places with a high rainfall. They are suited to the harsh, dry country, where Australian wool is produced. The wool is the finest in the world, but not very easy for hand spinners to handle. Merinos are kept mainly for wool-production, whereas these days the British breeds are mainly kept for meat, with wool as a bonus.

Backyarders want a quiet, friendly animal that gives a good wool crop, a fat lamb or two a year, and maybe some milk as well. The backyard sheep should be at home in the wetter areas of the country, for that is where many of the small stock farms are. For small areas and small flocks, most people prefer one of the old British breeds, or a cross. The fifty or so British breeds evolved over the centuries to cope with a great diversity of climate, and some of them are extremely tolerant of cold and wet weather.

Of the handful of breeds you meet in Australia, most will settle down in a small flock; it is only a matter of what they get used to. The following breeds are quite common, and often crossed with Merinos to produce lambs for meat.

Border Leicester   This is a big sheep with a roman nose, used for crossing in Australia and in Britain. It is early maturing, and mature ewes weigh 80–85 kg. The wool is long, 20 cm, which makes it easy to spin by hand.

Cheviot   This is a Scottish hill breed, the only one to find favour in Australia. These ancient sheep are small, with light bones. It was bred with bigger English sheep in the eighteenth century, and it is now a medium-sized, hardy sheep, used for crossing. In New Zealand, the shortwool Cheviot was crossed with the longwool Romney to produce the Perendale, a useful animal, which will give you good lambs and wool suitable for hand spinning.

Comeback   As the name implies, the Comeback is a first-cross ewe mated back to a Merino ram. The wool is a little too fine for hand spinning unless you are an expert. The Border Leicester is often used to breed the Comeback.

Corriedale   This sheep was bred in Australia and New Zealand by crossing the Lincoln, a large old British breed, with the Merino. The wool is fine and soft, longer than Merino wool. It is a good, dual-purpose breed, for wool and meat.

South Down The South Down is very popular in New Zealand for fat lambs.

Crossbred This is the name given to a cross between a Merino and one of the British breeds. It makes for a hardier animal, less likely to have diseases like foot rot, and more tolerant of wet weather. The Crossbred will produce a good crop of wool and can be mated with, say, a Suffolk ram to produce fat lambs for meat.

Polwarth Another Merino–Lincoln cross, the Polwarth is just one-quarter Lincoln and is quite popular for commercial production. They are mainly wool-producers, but without the deep neck folds of a typical Merino.

Romney Marsh This is a very solid sheep, originally bred in marshland, as the name suggests. It is one of the old 'longwool' breeds, but with a denser and finer wool. The staple length is about 15 cm and the fleece will weigh about 4 kg.

Shropshire and Hampshire Down These are similar to the Suffolk, and the rams are used to cross for fat lamb production.

Coloured sheep These can be of any breed, but are usually related, even if distantly, to the British breeds. Commercially, they used to be a nuisance and were culled from a commercial flock. But coloured wool is wonderful for hand spinning, because it looks interesting without being dyed. That is why nowadays coloured sheep are popular with small flock-owners, who may specialise in breeding coloured sheep. The gene for black is recessive, but if you mate two coloured sheep, they will produce coloured offspring.

Suffolk A compact sheep with black face and legs, this breed is very popular all over the world for lean meat production. The fleece is close and short, about 6.5 cm, but spinners say that the wool 'lacks character'. The Suffolk are more common in high rainfall areas of Australia.

Keeping sheep

It is wise to start small, and to make careful preparations. With sheep, fencing is the most important thing to start with. Sheep can wreck your garden and, even worse, your neighbours' gardens. So the first job is to make sure they will stay where you put them. Plain wire is best, because barbed-wire will damage the fleece. But if you have barbed-wire, don't go to the expense of changing it: most sheep live with it quite happily. The best, but most expensive type of sheep fencing is fabricated mesh, like the sort used for pigs. Plain wire fences with timber posts are the most common. This type will hold sheep if it is tight. You can put wire netting on the bottom two feet or so to keep small lambs from straying.

How will you catch the sheep again once you have let them loose in the paddock? Sheep dogs are essential for large flocks, and most commercial sheep-farmers have at least one good dog. But many people with small flocks of sheep can handle them easily without a dog. The secret is to train them to the bucket.

It will be best to keep your new sheep in a small area at first. This is where the training can take place. Give them some really palatable food in a bucket, like cattle concentrates or something with molasses in it. Once they come to expect it, let them follow you for a little way and then reward them with a feed. Gradually they will get used to the idea that when they follow you, food will be the reward. Then you will be able to move them anywhere, following the feed bucket. Training makes all animals easier to manage. Ours follow us from one paddock to another, because they are trained to follow when we call 'come on'. Usually there is a nice stretch of fresh grass at the end of it, so they are quite keen. The reward is important: fool them too often and they won't follow you any more.

A shed is useful, even though sheep do not usually need to be kept inside. At lambing time it is handy to put a ewe inside if she has a difficult lambing or if the weather is extreme. You can also store hay in it for the winter and maybe fleeces before you use or sell them.

Breeding

If you want to grow fibre and not produce meat, buy a few castrated males (wethers), and you will be able to keep them for years, taking a crop of wool every year with few problems. Don't keep an uncastrated ram, because he will become aggressive and unmanageable. Rams need special treatment, and I think it is better to run your ewes with a neighbour's flock to get them in lamb than to keep a ram for a small flock. You would need about thirty ewes before a ram becomes worthwhile. Ram lambs can be castrated and reared for the freezer, or sold if you can't face eating home-reared meat.

Vegetarian families can keep sheep for their wool and valuable manure, which

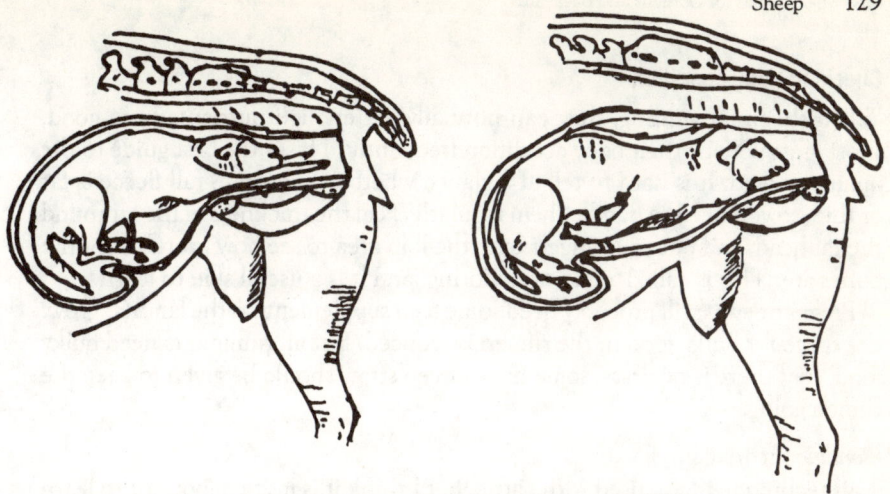

Lamb about to be born In the first diagram, the lamb is in the normal position for birth with forefeet presented first and nose on legs. The second diagram shows a problem presentation. It is best to get the vet if you have problems. You can learn as time goes by, so that eventually you will be able to correct simple problems yourself. Never be afraid to ask for help. Don't wait, or it may be too late. Remember that there may be more than one lamb waiting to be born.

helps to improve soil fertility. However, breeding is more fun, so you may like to start with females: ewe lambs, or older ewes in lamb, which will give you a lamb as well as a wool crop. Local farmers or local markets are possible sources of sheep. If you are in doubt, ask the stock and station agent to find you some healthy specimens. The breed you choose may determine where you go to look for sheep. Ewe lambs can be kept for breeding if you want to increase the flock.

Feeding

Grass is the usual food for sheep, and they will graze all the year round. An acre of good grass will support four or five sheep during the grazing season. However, you can't make rigid judgments about this, because every season is different. In a very good season, most people make surplus grass into hay.

In the cold winter months the high rainfall areas may not produce much grass, and some hay will be needed. Each sheep needs about a bale of hay a week. It is easy to calculate your needs: estimate the length of the winter period or the summer drought. Three months' supply of hay is always a good standby, say a hundred bales per head for your flock, allowing extra for pregnant and milking ewes.

Sheep also enjoy browsing, although they are not quite as keen as goats. Some of the new ideas for sustainable farming include providing browse for winter or drought feed instead of hay, since hay takes a lot of energy to produce. With this in mind, you could plant some areas of fodder tree, such as native casuarinas or acacias, or perhaps an introduced fodder tree such as tagasaste (tree lucerne) if it grows well in your area. The trees could be protected by netting so that the sheep nibble only the tips as they grow.

Checking body condition

When sheep are grazing, they can normally get enough nutrients from good, short grass. Check their body condition frequently. This is the best guide to the quality of feed. It is hard to tell at a glance whether a sheep in full fleece is fat or thin, so you need to handle them regularly. Feel the thickness of the fat round the tailhead, and put your fingers over the loin area to see how protuberant the bones are. This is called 'condition scoring' and it is a useful skill to learn.

Pregnant ewes will probably need some feed supplement: as the lamb(s) grow, the space for bulky food in the rumen is reduced. But all ruminants need bulky food, so if there is no grass, some hay or even straw should be given to keep the rumen going.

Condition scoring sheep

This technique is also used with cattle, but I think it is more important to learn to score sheep because their condition is harder to assess by eye, unless they have just been sheared. Condition scoring is for all sheep, not just for those used for meat. It is important that pregnant ewes are not too thin or too fat. Scoring can also be used to assess lambs for meat. The type of lamb you buy these days is a lean score 3, but if you like a fattier lamb, take it up to 4.

The aim is to regulate the feed so that sheep are never so thin as 1, not often so thin as score 2. Most sheep should be between 3 and 4, with 5 rather too fat for breeding. Don't worry about your scoring compared with other people's, but be consistent. If one of your sheep is much thinner than the others on the same regime, try to work out the reason. It may need veterinary attention.

First of all, make sure the sheep is standing level and square on its feet. Hold the head under the chin with one hand while you score with the other. In the centre of the back, behind the last rib and in front of the hip bone, are the spinous processes. Feel these and decide:
- Are the tips of the spinous processes sharp or rounded?
- Is the ridge of the spine raised above the level of muscle and fat cover?
- Does the muscle and fat cover rise above the ridge of the spine, leaving the spine in a slight depression?

Then feel for the transverse processes, coming out laterally from the spine, and decide:
- Do the bones feel sharp or smoothly rounded?
- How far will the tips of your fingers go under the transverse processes?
- Lastly, span the back with your hand and feel the overall thickness of muscle and fat on either side of the spine and between the spinous process and the transverse process.

Condition score 0 Very thin, no muscle or fat between skin and bone. This sheep is nearly dead, and I hope you never meet a sheep with a score of 0. If you do, it will need veterinary attention unless you are an experienced shepherd.

Condition score 1 Spinous processes are sharp and stick up. Transverse processes are sharp, and your fingers go under them easily. There is a hollow between the end of each process. Loin muscles are shallow, with no fat cover.

Condition score 2 Spinous processes are less sharp. With a little pressure you can push your fingers under the transverse processes. Loin muscles have more depth, but little fat cover.

Condition score 3 Spinous processes stick up only slightly and are smooth and rounded. Firm pressure is needed to feel them separately. Transverse processes are well covered and feel smooth. Firm pressure is needed to push your fingers under the ends. Loin muscles have some fat cover.

Condition score 4 Spinous processes can only be felt with firm pressure as a hard line, level with the flesh on either side. The ends of the transverse processes cannot be felt. Loin muscles are full, and well covered in fat.

Condition score 5 Bones cannot be felt at all. There is a hollow between the layers of fat on either side of the spine. Loin muscles have a thick fat cover.

Sheep care

For any of the routine tasks that involve handling the sheep, bring them quietly into a pen or yard and leave them for a while to settle down before starting work.

Foot trimming

Two or three times a year you will need to inspect the sheep's feet and trim them if necessary. The outer hoof grows, and on hard ground it wears down. But if the grass is soft or wet for long periods, the hoof will need trimming. If sheep are limping or grazing on their knees, an immediate inspection is called for.

To restrain a sheep for foot inspection, hold it in a sitting position against your legs, balanced on its rear end. Put the weight on the left side when working on the left hind leg, to prevent kicking. Transfer the weight to the right when you work on the right hind leg. First, brush the feet to remove the dirt so that you can see clearly. Look for infection (foot rot), which may be under an area of overgrown horn or may be an inflamed area between the claws. Using secateurs and a sharp clean knife, pare away all the overgrown horn. Secateurs will be needed for the hard outer hoof. Make the cutting strokes away from the sheep's body. Do not cut too deeply.

If the foot is infected, get expert help. Treatment may include formalin footbaths, aerosol sprays or injections, but a proper diagnosis will be needed. Until you have seen the various conditions that can afflict the feet of sheep, you will need a vet or an experienced shepherd to help you. After trimming you can drive the sheep through a footbath as a precautionary measure. A five per cent formalin solution is mixed up in a shallow tray with a ridged bottom. Ensure that there is about 5 cm of solution in the bath, that each animal has the feet immersed for at least a minute, and that the hooves are opened while in the bath. Then leave the sheep on concrete or clean straw for at least an hour before returning them to pasture. After foot trimming, sweep up the parings and burn them. Then rinse the instruments in disinfectant.

The conventional and most comfortable way to hold a sheep for inspecting the feet is like this. The sheep is gently tipped back onto its rump, to lean against your leg. In this position it is virtually immobile. You can hold it with the elbow, leaving your hands free to trim the feet. This is also the right position for shearing.

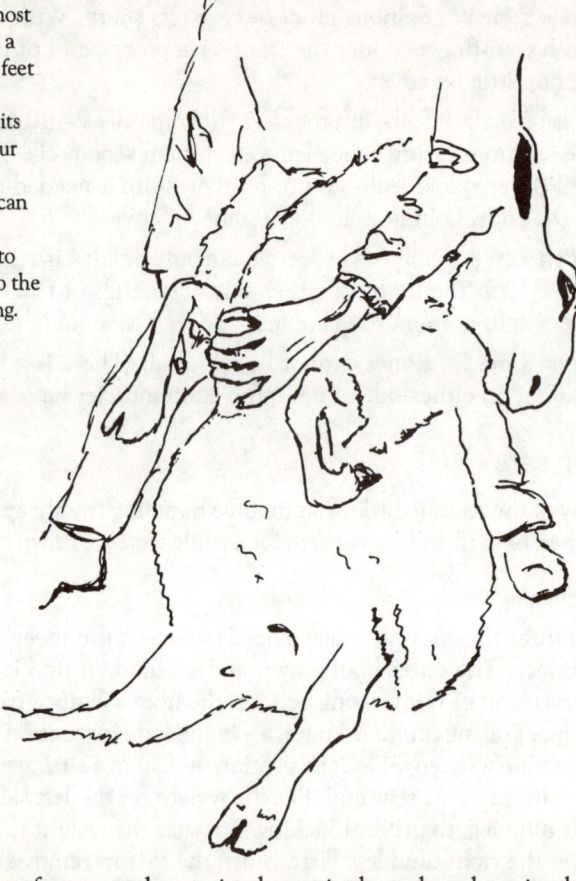

Routine medicines

Organic producers prefer not to dose animals routinely, unless there is a history of problems in the area. One common preventive routine is to vaccinate with a five-in-one vaccine against clostridial diseases. These are so called because the germs that cause them are clostridium organisms, which form resistant spores and live in the soil for long periods.

Pulpy kidney, blackleg and tetanus, malignant oedema and back disease are the five dangers against which the vaccine will protect the sheep, and it may be wise to use it if there is a history of clostridial diseases on neighbouring farms. The vaccination program consists of vaccinating ewes during the last month of pregnancy every year, which will give temporary immunity to the new-born lambs. The lambs are injected after a few days.

Worm drenches

With worms in mind, the aim is to move the sheep round onto fresh grass to prevent a worm build-up. Commercial sheep-farmers have a regular program of dosing for worms. On a small unit, with mixed stock, careful attention to health may mean that you can get away with using anthelmintics (wormers) only when a sheep shows symptoms of a worm problem, such as scouring, poor growth, loss of appetite or coughing.

It is, of course, natural for animals to have some parasites. When animals are in good health, with plenty of food, a balance can be struck and the parasites will not cause trouble to the animal. However, stress can alter a healthy balance, and then the worms will cause ill health. Organic producers try to keep the infestation down to as low a level as possible, but it is not possible to eliminate worms entirely.

Ewes at lambing and young lambs are probably the most at risk from worms. It is a good idea to put them, after lambing, in a paddock that has not had sheep on it for some time. This is why the best plan is to have several different classes of stock and some crops as well, so that the land gets a variety of uses.

External parasites

Parasites, such as lice, mites and blowflies, which lay eggs on the sheep, can be a real danger to sheep health. Flystrike in particular is most unpleasant: the maggots hatch out and proceed to eat the sheep wherever there is a wound or a patch of wet and dirty skin for them to gain entry. Dipping used to be the answer, a messy business of total immersion in disinfectant. You can dip your small flock in a tin bath on the lawn, but dipping is not compulsory in Australia.

The organophosphorus compounds used in modern sheep dips have several disadvantages. They can cause illness in the human operators, pollute water and leave residues in the meat. And, after all that, Australian blowflies have become resistant to them. A better approach is probably the pour-on compounds Cypor and Vetrazin, which are just poured along the sheep's back. These are based on cyromazine, which is not very toxic to mammals, but works against insects by acting as a growth regulator. These products will give the animal from six to eight weeks' protection.

Flystrike can sometimes be prevented by dagging. Dagging means cutting away the wool round the tail area, if it gets moist and dirty, so there is less to attract the fly to the sheep. Flies can attack the feet, head, back, almost anywhere on the sheep, and a watch should be kept for this horrible condition. When you find an infected animal, act immediately. Clip the wool round the area, comb out the maggots and treat with a disinfectant, preferably iodine based.

Shearing

It takes about three consecutive days of hard work to learn to shear a sheep, as I found when I arranged training courses for would-be shearers. Most small producers arrange for shearing to be done with another flock and find a competent shearer to do the job. It is best to shear when the weather gets warm. This is the natural time for the fleece to be shed. In warm weather there is less stress for the sheep, and they will not need as much energy to keep warm as they do in winter. So shearing is a job for November onwards, but before the weather gets too hot.

Worms

Worm-farming is the ultimate in environmental friendliness. Worms turn waste into valuable organic fertiliser, and worms can be let loose on organic farmers' land as well as being bred to provide food for zoo animals and fish. Earthworms can help increase food-production by improving the quality of the soil. Worms increase the water-holding capacity of the soil by improving soil structure, and they increase the micro-organisms in the soil. They also produce vermicompost or worm-castings; that is, digested material that provides plant nutrients. On a small holding, worms can be integrated with rabbits, fish farming and compost.

I have always thought that the American way of feeding dried sewage sludge to earthworms is a lot better than piping the stuff out to sea. Using sludge is illegal in Australia, so we use farm manure. Several successful worm farmers are using piggery waste as a raw material.

Species

Eisenia fetida, the manure worm, seems to be the best for producing compost in southern Australia. This is one of the introduced European species. It is also called Tiger worm or 'red wriggler' and one advertisement refers to the worms of this species as 'God's little helpers'. It is a fast-growing worm: its population can increase from 8 to 1500 in six months, given ideal conditions. However, the Tiger worm is not suitable for gardens or paddocks, because it needs a large amount of organic matter to be successful. This worm is often sold for fish bait.

Lumbricus rubellus is another temperate climate worm, although not as fast a breeder as the Tiger worm.

The earthworms of tropical Australia came from Africa, central America and South-East Asia. They include species such as *Pontoscolex corethrunus* and a group called *Pheretima*.

Aporrectodia caliginosa, a smallish worm with white lumps on the underside, is the most common in gardens and grassland in southern Australia. This is probably the one you should cultivate if you are growing worms to set free in the soil.

Housing

As a rule, worms are kept in special beds with a concrete base, or in bins with drainage at the bottom. I have seen successful small worm farms in polystyrene boxes, the sort fruit is packed in. (These boxes have a variety of uses as seedbeds,

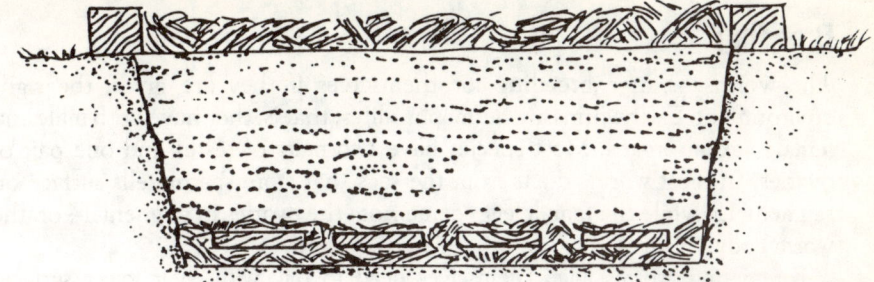

Earthworms in tropical regions In hot climates it is preferable to make a pit in which to keep the worms, rather than a raised bed. This is also a good idea in places where there are winter frosts. A shady place under trees will be good, perhaps in the orchard. Two metres is a good width. Old boards can be used to line the pit; the worms can hide under them when it gets really hot. A layer of straw covers the boards and on top of this is a mixture of manure and soil or sand, say three parts of rabbit manure to one of soil. Keep the mixture moist; the straw mulch helps to retain the moisture. In the wet season, protection from too much rain is needed. If the worms leave, conditions are not to their liking. Is the bed too hot or too cold, too wet or too dry? Have they enough food and is the pH right for them?

plant-tube-holders, and so on, and you can pick them up free from supermarkets.) The bed or box should be about 30 cm deep.

Metal containers are not suitable for worm-farming, because the dissolved metals in the compost could kill plants. If you want to go bigger than a fruit box, the next stage is to make raised beds about one metre wide. The concrete base makes it easier to collect the compost. Drainage holes will be needed if the beds are out in the open and exposed to rain, because worms don't like waterlogged conditions. Once the container is ready, the bed can be made. A light, porous material is essential because worms must breathe.

The worm-bed should be sited in a sheltered spot, out of direct sun, which could raise the temperature too high. Shade cloth will be needed if you have no suitable place with natural shade. A big tree would make a good shelter and would also keep out the rain. Worms are killed by frost, and slowed down by low temperatures, which is why commercial worm farmers operate in sheds.

Many people seem to mix manure, sometimes of different species (sheep, horse and cow manure are the best), with paper put through a hammer mill. The manure is allowed to cool down and rot for about a month before use, and the paper is wetted for about a week. Fresh sheep manure is too strong and should be leached. If no manure is available, grass cuttings can be used, lightened with straw, leaf mould or paper. Whatever you use, moisten the worm-bed so that a few drops of water can be squeezed from a handful of material.

When the bed is ready, the worms are introduced at the rate of 50 worms per litre of bedding, say 500 per fruit box or equivalent size. Bury worm food at intervals and then place the worms above it. Worms soon disappear, as they do in the garden, out of the light. Cover the bed with hessian or weedmat, or any other material that does not exclude air.

Breeding

The worms arrange breeding for themselves if they are given the right environment. Earthworms are hermaphrodites; that is, they have both male and female sexual organs. Most species have two pairs of testes and one pair of ovaries. In most worms ducts from the sacs that store sperm cells surface on segment 15, while the female eggs come from the ovaries in segment 14 on the worm body.

For mating, worms align themselves head to tail, with their lower surfaces touching. Each receives the other's sperm and then lays eggs in a ring formed round the body, which passes over the stored sperm as the worm wriggles out of it. The ring is shed and forms a cocoon round the fertilised eggs, which hatch out in a few weeks, the exact time depending on the temperature. (Worms breed faster in a warm bed.) Each cocoon gives rise to only one or two worms, but the Tiger worm may produce several hundred cocoons in a year. The young worms reach maturity in about six months and have a potential life span of several years.

Feeding

Worms need organic material with at least 1 per cent nitrogen, which gives you a wide range of choice. Plant material usually contains more than this and can be mixed with paper or cardboard. The cellulose and the bacteria in animal manures are good food for worms. They can eat about half their own weight in food every day.

Like poultry, earthworms need a little grit to help grind the food, so a little soil can be added to provide it. Many people feed the worms on kitchen scraps, but if you give them too much, the food will sour and the smell will be terrible. If the environment is right, the worms will stay and breed happily. But – an awful warning – if the worms do not like what you provide they will leave, migrating out of the beds or pits in search of a better home. Worms will leave if the bed is too hot or too cold, too acid or too alkaline, too wet or too dry. Worms prefer a temperature range of 17–25°C, and a pH of 6.5–6.8 (they cannot tolerate acid soils). Check the pH and sprinkle with dolomite (naturally occurring limestone), if the bed is too acid. Wood ash can also be used. Good management will keep your worms at home.

The beds should be damp, but not soggy. Indoor beds will need to be sprinkled with water about once a week, more in warm, dry weather. If white worms appear, the bed is probably too wet. If a bed smells unpleasant, it is not run properly, and it may attract flies. The beds should give off a pleasant, earthy odour.

Harvesting the vermicompost

In about six months, your first batch of compost should be ready. To get really high-quality castings, people stop feeding the worms a few weeks before harvesting, so that they work through the bed again. Most vermicompost is a mixture of worm-castings and bed material. The worm-castings are a very good

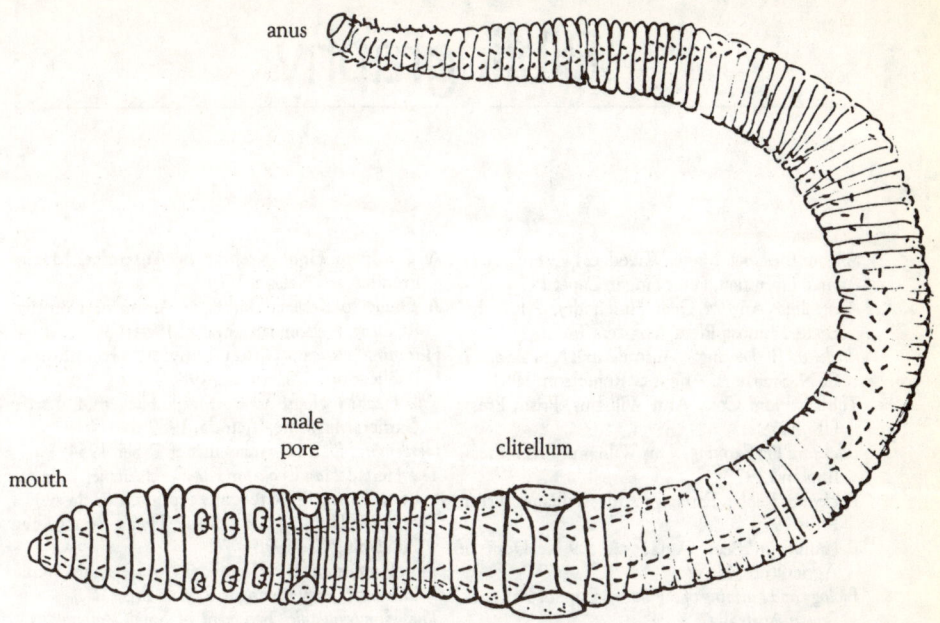

The common European earthworm is of the *Lumbricides* family, which can be found in temperate regions all over the world. The earthworm used to produce vermicompost from manure won't survive long in ordinary soil without the rich organic matter of the manure heap. To increase the earthworm population of your garden, you will need field worms or night crawlers (*Lumbricus terrestris*), which appear on the lawn at night.

plant food because the nutrients are more available than they were before being processed by the worms.

The worms can be taken out and put into a fresh bed, and the compost is a valuable fertiliser with, on average, about 2 per cent nitrogen, 1 per cent phosphorus, and 1 per cent potassium. Sixty thousand worms will produce a tonne of castings in about six months, and commercial worm-farmers usually reckon their stock in millions.

Harvesting is hard work, and it seems that many worm farmers get tired of sorting worms by hand and opt out of the industry. There are several techniques for separating the worms from the compost. If the bed is piled into a heap under a bright light, the worms will move down into the heap and the top can be skimmed off as compost. Some people invent sorting machines. These mainly consist of rotating drums, which throw the compost to the outside – since it is lighter than the worms – where it hits a screen and is separated from the worms.

As worm-farming is seen to work, the time may come when every garden has its own worm-farm in the corner. It seems such a sensible thing to do. And if you are concerned about the manure from your animals on a small block, worms will help to convert it to compost so quickly that there will be no pollution problem.

Bibliography

Angora Breeding, Mavis Walledge, Nelson, 1987
Animal Liberation, Peter Singer, Cape, 1976
Australian Angora Goat Husbandry, Alma M. Bode, Mimosa Press, Charters Towers
Backyard Beekeeping in Australia and New Zealand, C. N. Smithers, Angus & Robertson, 1987
The Backyard Cow, Ann Williams, Prism Press, UK, 1979
Backyard Pig Farming, Ann Williams, Prism Press, England 1977
Backyard Rabbit Farming, Ann Williams, Prism Press 1978
Beginning in Bees, Agfact A 8.9.1., Dept of Agriculture, NSW
Biology and Farming of the Yabby, Dept of Fisheries, South Australia
Black's Veterinary Dictionary, A&C Black, 15th edition, 1985
Common Sense with Horses, Agfact A 6.2.3, Dept of Agriculture, NSW
The Complete Training of Horse and Rider, A. Podhajsky, Harrap, 1978
Dog Care Question and Answer, Barry Bush, Orbis, 1982
Dog Training: the Gentle Modern Method, David Weston, Hyland House, Melbourne
Earth Garden Magazine (6 issues a year), editor Alan Gray, RMB 427 Trentham, Vic. 3458
The Earthworm Book, Jerry Minnich, Rodale Press, USA, 1977
Earthworms for Gardeners and Fishermen, Handreck & Lee, CSIRO, Melbourne, 1986
Earthworms in Australia, Hyland House, Melbourne, 1993
The Family Cow, Dirk van Loon, Garden Way Publishing, 1976
The Family Smallholding, Katie Thear, Batsford, 1983
Farming in a Small Way, NSW Dept of Agriculture
The Farming of Goats for Fibre and Meat in New Zealand, David Yerex, Ampersand, 1986
Fish for Farm Dams, Ric Fallu, Dept of Agriculture, Vic. (toll free number 008 800 755)
Fish in Farm Dams, Agfact F3.1.1, NSW Dept of Agriculture
The Goat Manual, Dept of Agriculture, NSW, PO Box H220, Haymarket, NSW 2000
Grass Roots Magazine (6 issues a year), editors Megg Miller & Mary Horsfall, Box 242 Euroa, Vic. 3666
A Guide to Goat Keeping in Australia, Maria Prendergast, Nelson, 1981
A Guide to Keeping Poultry in Australia, Dorothy Reading, Nelson, Melbourne, 1984
Handling Horses, Garda Langley, Greenhouse Publications, Melbourne, 1988
The Healthy House Cow, Marja Fitzgerald, Earth Garden Magazine, Victoria, 1989
Herdsmanship, F. Newman Turner, Faber, 1954
The Herbal Handbook for Farm and Stable, J. de B. Levy, Faber, 1973, recently reprinted (a classic).
The Home Dairying Book, Katie Thear, Broad Leys Publishing Co, 1978
The Homesteaders Handbook for Raising Small Livestock, Jerry Belanger, Rodale Press
The Homoeopathic Treatment of Small Animals, C. Day, Wigmore Publications, London, 1984
Horse Health: Practical Worm Control, Agfact A 6.9.5, Dept of Agriculture, NSW
Introduction to Permaculture, Bill Mollison, Tagari Publications, 1991
Making your own Cheese and other Dairy Products, Margaret Barca, Nelson, Melbourne, 1978
Management and Welfare of Farm Animals, Bailliere Tindall, 1988, Universities Federation of Animal Welfare (Handbook)
A Modern Herbal, Mrs M. Grieve, Penguin 1976 (first published 1931)
Natural Horse Care, Pat Coleby, Kangaroo Press, Sydney, 1989
The New Duck Handbook, Heinz-Sigurd Raethel, Barron's Educational Series, New York, 1988
Organic Farming, Nicolas Lampkin, Farming Press, England, 1990
Practical Feeding of Horses, Agfact A 6 5.3, Dept of Agriculture, NSW
Raising the Homestead Hog, Jerome D. Belanger, Rodedale Press, USA, 1977
Town and Country Farmer Magazine (4 issues a year), editors Glenn & Shirley Hurley, PO Box 798 Benalla, Vic. 3672
A Way of Life: Sheepdog Training and Handling, Glyn Jones & Barbara Collins, Farming Press, 1987
The Weekly Times, GPO Box 751F Melbourne, Vic. 3001
The Wilderness Garden, Jackie French, Aird Books, Melbourne, 1992

Index

acid curd cheese 34
agribusiness 3, 7
angora goats 34, 83
angora rabbits 35, 124

battery hen systems 7
behaviour of animals 2
Bighorn sheep 2
bloat
  in cows 53, 56
  in goats 90
butter 32

calf rearing 58-60
cashmere goats 34, 84
cheese 33, 34
coccidiosis 102
colic, in horses 108
comfrey
  animal health 14
  fodder crop 27
colouring agents
  annatto, in butter 32
  canola, in honey 45
condition scoring 130
cream 32

dam 22, 77
  ducks 69, 70
  fish 74-78
  geese 79
disease 13-15
  bees 44
  colic, in horses 108
  cows 55, 56
  dogs 67
  hens 101
  ketosis 54
  pigs 118
domestication 1, 2, 92
drugs 13

factory farms 3
  goats, not kept on 82

fertility 4, 23
  worms 134
fish
  for ducks 72
  for geese 79
fodder crops 25-27
foot care
  cows, infection 56
  goats 88
  horses 109
  sheep 131
free range chickens 96

handling of animals
  people, suitability 6, 8
  poor, and behaviour 2
  regulations 5, 17, 18
  rules 4
  shooting 16
  tethering 11
  welfare, five freedoms 7
health, see disease
honey 41, 49, 50

integration of species 20-23
intensive farming 7
  animal health, and 13
  goats, not practised 82

kids, goat 85

mastitis
  cows 55
  goats 87
milk fever 56
milk products, for poultry 72
mohair 82

native trees
  fodder crops 26
  honey 45

obesity in dogs 64
organic farming, society 21

paulownia trees 26
permaculture 18
pets
   therapeutic effects 1, 67
   unusual species 3
ponds 22
   ducks 69, 70
   geese 79
   fish 74-78
preserving meat 38
prevention, of disease 19

regulations for animal keeping 5, 17, 18
   fish 77
   rabbits 121
rumination 53

salting meat 38, 120
scotch hands 33
shooting animals, in emergency 16
slaughtering, at home 37
   pigs 119
sleepy cream 33

solar wax extractor 41
spinning 35, 36
starter culture, for dairy products 27
suckler cows 60
sustainable agriculture 4, 18
   national association for 21

tagasaste 25
tethering 11
therapeutic effects of animals on people 1, 67
training
   dogs 64
   sheep 128

vaccination 13
   dogs 67
vermicompost 136

wax extractor, solar 41
willows 27
WWOOF 3, 19

yoghurt 32